Knowledge Transfer in the Automobile Industry

The book arose from a multi-disciplinary study which looked at the development of global-local manufacturing clusters in the context of a developing, Asian economy. The study demonstrates the connection among theoretical perspectives such as international business, development studies, economic geography, and organisational learning clusters/production networks through an in-depth case study of the Indonesian automotive cluster. The book gives a detailed account of two automotive clusters (Toyota and Honda) and their contribution to regional economic development in emerging economies in the Asian region. The book builds on existing literature to develop a theoretical framework to shed light on the study's empirical findings.

The book discusses practical implications for both the business community and policymakers. The discussion on global-local networks in an Asian context supplements existing literature and case studies in the field. This is one of the few books that explicitly links regional clusters to global networks. The book offers a refreshingly international (Asian) perspective to the literature on clusters and economic geography for emerging economies.

Dessy Irawati, PhD, is a Teaching Associate at Newcastle University Business School in the UK and a Visiting Lecturer at the Graduate School of Management, University Putra Malaysia. She gained her PhD in International Business Strategy and Economic Geography in 2009 from Newcastle University Business School, UK. Her disciplinary background is in International Business Strategy, Economic Geography, and Regional Studies. She explores why some regions have a better economic performance than others and argues that this is because they encourage knowledge creation in the global-local networks more than other regions.

Furthermore, she has researched and taught international business management, investigating overlaps with the fields of strategy, organisation and learning. Alongside this, she continues to develop her research interests on innovation and regional development in the knowledge-based economy, specifically in the context of agglomeration, industries, and networks. Her extended research interests are: international business strategy, multinational enterprises (MNEs), small- and medium-sized enterprises (SMEs), innovation and technology management, globalisation and development studies, cluster-based policy and networks, industrial dynamics, and knowledge transfer.

Routledge studies in the modern world economy

1 **Interest Rates and Budget Deficits**
A study of the advanced economies
Kanhaya L. Gupta and Bakhtiar Moazzami

2 **World Trade after the Uruguay Round**
Prospects and policy options for the twenty-first century
Edited by Harald Sander and András Inotai

3 **The Flow Analysis of Labour Markets**
Edited by Ronald Schettkat

4 **Inflation and Unemployment**
Contributions to a new macroeconomic approach
Edited by Alvaro Cencini and Mauro Baranzini

5 **Macroeconomic Dimensions of Public Finance**
Essays in honour of Vito Tanzi
Edited by Mario I. Blejer and Teresa M. Ter-Minassian

6 **Fiscal Policy and Economic Reforms**
Essays in honour of Vito Tanzi
Edited by Mario I. Blejer and Teresa M. Ter-Minassian

7 **Competition Policy in the Global Economy**
Modalities for co-operation
Edited by Leonard Waverman, William S. Comanor and Akira Goto

8 **Working in the Macro Economy**
A study of the US labor market
Martin F. J. Prachowny

9 **How Does Privatization Work?**
Edited by Anthony Bennett

10 **The Economics and Politics of International Trade**
Freedom and trade: volume II
Edited by Gary Cook

11 **The Legal and Moral Aspects of International Trade**
Freedom and trade: volume III
Edited by Asif Qureshi, Hillel Steiner and Geraint Parry

12 **Capital Markets and Corporate Governance in Japan, Germany and the United States**
Organizational response to market inefficiencies
Helmut M. Dietl

13 **Competition and Trade Policies**
Coherence or conflict
Edited by Einar Hope

14 **Rice**
The primary commodity
A. J. H. Latham

15 **Trade, Theory and Econometrics**
Essays in honour of John S. Chipman
Edited by James C. Moore, Raymond Riezman and James R. Melvin

16 **Who benefits from Privatisation?**
Edited by Moazzem Hossain and Justin Malbon

17 **Towards a Fair Global Labour Market**
Avoiding the new slave trade
Ozay Mehmet, Errol Mendes and Robert Sinding

18 **Models of Futures Markets**
Edited by Barry Goss

19 **Venture Capital Investment**
An agency analysis of UK practice
Gavin C. Reid

20 **Macroeconomic Forecasting**
A sociological appraisal
Robert Evans

21 **Multimedia and Regional Economic Restructuring**
Edited by Hans-Joachim Braczyk, Gerhard Fuchs and Hans-Georg Wolf

22 **The New Industrial Geography**
Regions, regulation and institutions
Edited by Trevor J. Barnes and Meric S. Gertler

23 **The Employment Impact of Innovation**
Evidence and policy
Edited by Marco Vivarelli and Mario Pianta

24 **International Health Care Reform**
A legal, economic and political analysis
Colleen Flood

25 **Competition Policy Analysis**
Edited by Einar Hope

26 **Culture and Enterprise**
The development, representation and morality of business
Don Lavoie and Emily Chamlee-Wright

27 **Global Financial Crises and Reforms**
Cases and caveats
B. N. Ghosh

28 **Geography of Production and Economic Integration**
Miroslav N. Jovanović

29 **Technology, Trade and Growth in OECD Countries**
Does specialisation matter?
Valentina Meliciani

30 **Post-Industrial Labour Markets**
Profiles of North America and Scandinavia
Edited by Thomas P. Boje and Bengt Furaker

31 **Capital Flows without Crisis**
Reconciling capital mobility and economic stability
Edited by Dipak Dasgupta, Marc Uzan and Dominic Wilson

32 **International Trade and National Welfare**
    Murray C. Kemp

33 **Global Trading Systems at Crossroads**
    A post-Seattle perspective
    Dilip K. Das

34 **The Economics and Management of Technological Diversification**
    Edited by John Cantwell, Alfonso Gambardella and Ove Granstrand

35 **Before and Beyond EMU**
    Historical lessons and future prospects
    Edited by Patrick Crowley

36 **Fiscal Decentralization**
    Ehtisham Ahmad and Vito Tanzi

37 **Regionalisation of Globalised Innovation**
    Locations for advanced industrial development and disparities in participation
    Edited by Ulrich Hilpert

38 **Gold and the Modern World Economy**
    Edited by MoonJoong Tcha

39 **Global Economic Institutions**
    Willem Molle

40 **Global Governance and Financial Crises**
    Edited by Meghnad Desai and Yahia Said

41 **Linking Local and Global Economies**
    The ties that bind
    Edited by Carlo Pietrobelli and Arni Sverrisson

42 **Tax Systems and Tax Reforms in Europe**
    Edited by Luigi Bernardi and Paola Profeta

43 **Trade Liberalization and APEC**
    Edited by Jiro Okamoto

44 **Fiscal Deficits in the Pacific Region**
    Edited by Akira Kohsaka

45 **Financial Globalization and the Emerging Market Economies**
    Dilip K. Das

46 **International Labor Mobility**
    Unemployment and increasing returns to scale
    Bharati Basu

47 **Good Governance in the Era of Global Neoliberalism**
    Conflict and depolitization in Latin America, Eastern Europe, Asia and Africa
    Edited by Jolle Demmers, Alex E. Fernández Jilberto and Barbara Hogenboom

48 **The International Trade System**
    Alice Landau

49 **International Perspectives on Temporary Work and Workers**
    Edited by John Burgess and Julia Connell

50 **Working Time and Workers' Preferences in Industrialized Countries**
    Finding the balance
    Edited by Jon C. Messenger

51 **Tax Systems and Tax Reforms in New EU Members**
    Edited by Luigi Bernardi, Mark Chandler and Luca Gandullia

52 **Globalization and the Nation State**
The impact of the IMF and the World Bank
Edited by Gustav Ranis, James Vreeland and Stephen Kosak

53 **Macroeconomic Policies and Poverty Reduction**
Edited by Ashoka Mody and Catherine Pattillo

54 **Regional Monetary Policy**
Carlos J. Rodríguez-Fuentez

55 **Trade and Migration in the Modern World**
Carl Mosk

56 **Globalisation and the Labour Market**
Trade, technology and less-skilled workers in Europe and the United States
Edited by Robert Anderton, Paul Brenton and John Whalley

57 **Financial Crises**
Socio-economic causes and institutional context
Brenda Spotton Visano

58 **Globalization and Self Determination**
Is the nation-state under siege?
Edited by David R. Cameron, Gustav Ranis and Annalisa Zinn

59 **Developing Countries and the Doha Development Round of the WTO**
Edited by Pitou van Dijck and Gerrit Faber

60 **Immigrant Enterprise in Europe and the USA**
Prodromos Panayiotopoulos

61 **Solving the Riddle of Globalization and Development**
Edited by Manuel Agosín, David Bloom, George Chapelier and Jagdish Saigal

62 **Foreign Direct Investment and the World Economy**
Ashoka Mody

63 **The World Economy**
A global analysis
Horst Siebert

64 **Production Organizations in Japanese Economic Development**
Edited by Tetsuji Okazaki

65 **The Economics of Language**
International analyses
Edited by Barry R. Chiswick and Paul W. Miller

66 **Street Entrepreneurs**
People, place and politics in local and global perspective
Edited by John Cross and Alfonso Morales

67 **Global Challenges and Local Responses**
The East Asian experience
Edited by Jang-Sup Shin

68 **Globalization and Regional Integration**
The origins, development and impact of the single European aviation market
Alan Dobson

69 **Russia Moves into the Global Economy**
John M. Letiche

70 **The European Economy in an American Mirror**
Barry Eichengreen, Michael Landesmann and Dieter Stiefel

71 **Working Time Around the World**
Trends in working hours, laws, and policies in a global comparative perspective
Jon C. Messenger, Sangheon Lee and Deidre McCann

72 **International Water Treaties**
Negotiation and cooperation along transboundary rivers
Shlomi Dinar

73 **Economic Integration in the Americas**
Edited by Joseph A. McKinney and H. Stephen Gardner

74 **Expanding Frontiers of Global Trade Rules**
The political economy dynamics of the international trading system
Nitya Nanda

75 **The Macroeconomics of Global Imbalances**
European and Asian perspectives
Edited by Marc Uzan

76 **China and Asia**
Economic and financial interactions
Edited by Yin-Wong Cheung and Kar-Yiu Wong

77 **Regional Inequality in China**
Trends, explanations and policy responses
Edited by Shenggen Fan, Ravi Kanbur and Xiaobo Zhang

78 **Governing Rapid Growth in China**
Equity and institutions
Edited by Ravi Kanbur and Xiaobo Zhang

79 **The Indonesian Labour Market**
Shafiq Dhanani, Iyanatul Islam and Anis Chowdhury

80 **Cost-Benefit Analysis in Multi-level Government in Europe and the USA**
The case of EU cohesion policy and of US federal investment policies
Alessandro Ferrara

81 **The Economic Geography of Air Transportation**
Space, time, and the freedom of the sky
John Bowen

82 **Cartelization, Antitrust and Globalization in the US and Europe**
Mark LeClair

83 **The Political Economy of Integration**
Jeffrey Cason

84 **Critical Issues in Air Transport Economics and Business**
Rosario Macario and Eddy Van de Voorde

85 **Financial Liberalisation and Economic Performance**
Luiz Fernando de Paula

86 **A General Theory of Institutional Change**
Shiping Tang

87 **The Dynamics of Asian Financial Integration**
Edited by Michael Devereux, Philip Lane, Park Cyn-young and Wei Shang-jin

88 **Innovative Fiscal Policy and Economic Development in Transition Economies**
Aleksandr Gevorkyan

89 **Foreign Direct Investments in Asia**
Edited by Chalongphob Sussangkarn, Yung Chul Park and Sung Jin Kang

90 **Time Zones, Communications Networks, and International Trade**
Toru Kikuchi

91 **Miraculous Growth and Stagnation in Post-War Japan**
Edited by Koichi Hamada, Keijiro Otsuka, Gustav Ranis, and Ken Togo

92 **Multilateralism and Regionalism in Global Economic Governance**
Trade, investment and finance
Edited by Junji Nakagawa

93 **Economic Growth and Income Inequality in China, India and Singapore**
Trends and policy implications
Pundarik Mukhopadhaya, G. Shantakumar and Bhanoji Rao

94 **Foreign Direct Investment in China**
Spillover Effects on Domestic Enterprises
Deng Ziliang

95 **Enterprise Forms and Economic Efficiency**
Capitalist, cooperative and government firms
Kazuhiko Mikami

96 **Diversity and Transformations of Asian Capitalism**
Edited by Boyer, Uemura and Isogai

97 **Knowledge Transfer in the Automobile Industry**
Global-local production networks
Dessy Irawati

Knowledge Transfer in the Automobile Industry
Global-local production networks

Dessy Irawati

First published 2012
by Routledge
2 Park Square, Milton Park, Abingdon, Oxon OX14 4RN

Simultaneously published in the USA and Canada
by Routledge
711 Third Avenue, New York, NY 10017

Routledge is an imprint of the Taylor & Francis Group, an informa business

© 2012 Dessy Irawati

The right of Dessy Irawati to be identified as author of this work has been asserted by her in accordance with the Copyright, Designs and Patents Act 1988.

All rights reserved. No part of this book may be reprinted or reproduced or utilised in any form or by any electronic, mechanical, or other means, now known or hereafter invented, including photocopying and recording, or in any information storage or retrieval system, without permission in writing from the publishers.

Trademark notice: Product or corporate names may be trademarks or registered trademarks, and are used only for identification and explanation without intent to infringe.

British Library Cataloguing in Publication Data
A catalogue record for this book is available from the British Library

Library of Congress Cataloging in Publication Data
Irawati, Dessy.
 Knowledge transfer in the automobile industry : global-local production networks / Dessy Irawati.
 p. cm. – (Routledge Studies in the modern world economy ; 97)
 Includes bibliographical references and index.
 1. Automobile industry and trade–Indonesia. 2. Automobile industry and trade–Japan. 3. Industrial clusters–Indonesia. 4. Technology transfer–Japan. I. Title.
 HD9710.I5I73 2011
 629.2068'4–dc23
                                        2011020515

ISBN: 978-0-415-68060-8 (hbk)
ISBN: 978-0-203-69885-3 (ebk)

Typeset in Times New Roman
by Wearset Ltd, Boldon, Tyne and Wear

Printed and bound in Great Britain by
TJI Digital, Padstow, Cornwall

Contents

List of figures, maps and tables   xiv
Acknowledgements   xv

1 **Clusters and knowledge transfer into the Indonesian automotive industry**   1
  Research background and aims 2
  Research questions 3
  Book structure 5

2 **Theoretical framework on clusters**   7
  Cluster definition 7
  Clusters and competitive advantage 9
  Clusters and networks 12
  Proximity and industrial clusters: the impact on networks and technology 14
  Cluster boundaries 16
  Benefits from clusters 17
  Cluster analysis methods 20
  Clusters and policy 26
  Clusters in developing countries and FDI 37
  Conclusion: bringing the strands together 39

3 **Production and knowledge transfer in Japanese automotive networks**   43
  Knowledge transfer by Japanese MNEs in the automotive industry 44
  Japanese automotive production networks in Southeast Asia 48
  Vertical and horizontal networks in the automotive industry 51
  FDI in the Japanese automotive industry and its supply chain networks in Southeast Asia 53
  Conclusion 57

Contents

4 **Global and national environments: the macroeconomic context in Indonesia**   59
  The Indonesian macroeconomy after the Asian financial crisis 59
  Manufacturing investment in Indonesia 61
  Indonesian industrial policy in the automotive sector 63
  Cluster strategy in Indonesia: from SMEs to industry 70
  Conclusion 80

5 **Methodologies**   82
  Data analysis 97
  Research limitations 99
  Research ethics 100
  Conclusion 100

6 **The importance of the Java region for the Indonesian automotive cluster**   103
  The importance of Java as an industrial location 103
  The economic, spatial and cultural context in Java 107
  Automotive cluster characteristics in Java 109
  Organising capacity of the automotive cluster in Java 114
  Conclusion 118

7 **Car production in Indonesia: the Toyota complex**   119
  The Indonesian subsidiary within the Japanese MNE 119
  Knowledge transfer within the global keiretsu 125
  Knowledge transfer within the Indonesian automotive cluster 130
  Conclusion 143

8 **Motorcycle production in Indonesia: the Honda complex**   145
  The Indonesian subsidiary within the Japanese MNE 146
  The Indonesian automotive cluster supply chain 149
  Knowledge transfer within the Indonesian automotive cluster 154
  Conclusion 158

9 **The Indonesian automotive cluster in Toyota and Honda's global production networks**   160
  Toyota and Honda's significant presence in the development of the Indonesian automotive cluster 160
  Clusters in the context of the Indonesian automotive industry 164
  The learning aspect of knowledge networks 166
  Indonesian government policy and Japan 170
  Conclusion 173

10 Conclusion 176
The cluster concept 176
Theoretical contributions 177
Areas for future research 181

Notes 183
Bibliography 185
Index 204

Figures, maps and tables

Figures

1.1	Conceptual framework before fieldwork	4
5.1	Research design: research question 1	89
5.2	Research design: research question 2	91
5.3	Research design: research question 3	93
6.1	The automotive cluster in Java, Indonesia	109
7.1	Karawang plant layout	122
7.2	Simple model of the tiered supply chain	131
9.1	Research framework analysis before fieldwork	162
9.2	Research framework analysis after fieldwork	163

Maps

6.1	Automotive cluster in the Java region Indonesia	112
7.1	TMMIN's global market, 2007	124
7.2	Distribution of TMMIN's domestic suppliers, Java	139

Tables

6.1	Major carmakers, Java	110
6.2	Major motorcycle makers, Java	111
8.1	Honda's supply value chain network, Indonesia	153

Acknowledgements

I would like to express my gratitude to all those who gave me the possibility to complete this research monograph.

I want to thank Routledge for giving me the opportunity to disseminate my thoughts and research to readers worldwide. I really appreciate it.

I want to thank the Knowledge Innovation Technology and Enterprise (KITE) research centre at Newcastle University Business School for giving me a tremendous research place to commence this research project, to do the necessary research, and to use all the facilities and available training. I also want to show my appreciation to the Regional Studies Association for the stimulating research environment and the opportunity to share the ideas with another scholar.

Foremost, I would like to express my sincere gratitude to my primary supervisor, Prof. David Charles at Strathclyde University, Glasgow, for his continuous support of my PhD study and research, for his patience, motivation, enthusiasm, and immense knowledge. His guidance helped me throughout my research and the writing of this book. I could not have imagined having a better adviser and mentor for my PhD.

I would like to thank my thesis examiners, Prof. David Bailey at Birmingham University and Prof. Joanne Roberts at Newcastle Business School at Northumbria University, for their encouragement, insightful comments, and hard questions.

Additionally, I am deeply indebted to my best friend, Dr Joanna Berry, whose help, stimulating suggestions and encouragement helped me during the research for and writing of my thesis and book, and in my leisure time. Appreciation also to Charlie Berry, my best little buddy, who always makes me smile and happy.

My thesis and this book would not have been possible without the support and encouragement of my senior colleagues and friends in Newcastle University Business School of and my career adviser, David Levinson: thank you for all your guidance.

I cannot end without thanking my family in Indonesia and in the Netherlands, on whose constant encouragement and love I have relied throughout my time at Newcastle University Business School. Their unflinching courage and conviction will always inspire me, and I hope to continue, in my own small way, the noble mission to which they gave their lives. It is to them that I dedicate this work.

Last but not least, I would like to thank Dr Roel Rutten, Fellow at the Centre for Innovation Research and Assistant Professor in the Faculty of Social and Behavioural Science, Department of Organisation Studies at Tilburg University, Netherlands, whose patient love enabled me to complete this research monograph.

A journey has just begun and there are more discoveries to be made.

Dessy Irawati
Newcastle upon Tyne,
30 July 2011

1 Clusters and knowledge transfer into the Indonesian automotive industry

Clusters are said to have the discreet charm of ambiguous objects of desire (Malmberg *et al.* 1996; Steiner 1998; Malmberg and Maskell 2004). This charm lies in the idea that regional specialisation around the interlinked activities of complementary firms in production and service sectors and their cooperation with public, semi-public and private research and development (R&D) institutions creates synergies, increases productivity, and leads to economic advantages. Therefore, regions should specialise, and policy should generate, develop, and support such clusters (Benneworth and Charles 2001; Caniëls and Romijn 2005; Feldman *et al.* 2005a; Asheim and Coenen 2006).

In recent times, therefore, clusters have grown to be an object of aspiration for many regions wanting to advance their knowledge and technology (e.g. Baden-Württemberg, the Third Italy, Silicon Valley, Route 128 – Boston and Cambridge). They may be based on different foundations, as Pavitt (1987a) explains: the obscurity of clusters stems from this multidimensionality, which means that clusters are based on different economic dimensions, take different forms, are measured and quantified by relatively different methods and empirical approaches, and are legitimated by a range of theories and hypotheses. Additionally, they have also become a desirable object of research: the still vague character of clusters poses problems of a theoretically sound definition, of empirical measurement, of policy recommendation, and of evaluation (Morgan and Nauwelaers 1999; Maskell 2001).

Nevertheless, little is known about the critical success factors that determine the economic development of cities and regions, and empirical studies that draw lessons for policy are scarce (Tichy 1998; Nijkamp 1999; Ernst 2000; Romjin 2002; Metcalfe and Ramlogan 2006). There is a lack of cluster studies in the automotive sector for developing countries in the Southeast Asian region, specifically for Indonesia (Kuncoro 2002; Gammeltoft and Aminullah 2004). Due to this lack of cluster-based literature in Indonesia, there is a need to investigate the development of clusters in this country, particularly to what extent a sectoral view (in this instance, the automotive industry) and knowledge transfer in the cluster is adequate to analyse regional economic growth and cluster-based policies in Indonesia. There are many indications that industrial economic growth increasingly emerges from fruitful cooperation between multinational enterprises (MNEs) in a vibrant region like Java.

2 Clusters and knowledge transfer

Currently, based on data from the Indonesian Ministry of Industry and Trade (MOTIRI 2005), there are 40 industrial estates that have been labelled as clusters currently operating in Indonesia. Of these, 32 are located on the island of Java, the country's most densely populated region. Alongside its cluster strategy, the Indonesian government initiated the setting up of industrial districts, later known as industrial clusters, to provide a suitable location for both domestic and international investors by providing all necessary infrastructure, facilities and housing in one safe location – at a reasonable cost – thus providing a secure base for industry and manufacturing (OECD 1999b), and specifically for the automotive industry as a key contributor to the economy (MOTIRI 2005).

The clustering of firms in the automotive industry to provide complementary webs of activity in the same sector can improve a region's competitiveness and knowledge. Policies aimed at sustaining such clusters, thus promoting local-global cooperation, are necessary. Therefore, this research is aimed at global-local production, which accelerates the development of the Indonesian automotive cluster through the knowledge transfer processes between Toyota and Honda as the leading multinational enterprises (MNEs) and their local transplants and suppliers. The types and structures of relationships and networks covered by knowledge transfer in the automotive sector are complex because of the features of the knowledge base, the relevant learning processes, the basic technology, the characteristics of demand, the key links, and dynamic complementarities. Still, networks in the automotive sector are particularly significant with regard to innovation. Strong international competition and fast technological development push firms to constantly innovate in terms of products or services, processes, and markets. Consequently, involvement in a network helps a firm to focus on core capabilities, and provides access to resources (specific knowledge, technology, financial means, product, assets, market) in another organisation (Ernst 2000; Saxenian 2005; Breschi and Malerba 2005). This supports improvements to their competitive position.

The object of this research is the process of knowledge transfer in the automotive clusters in Indonesia, specifically looking at the case studies of Toyota and Honda, the leading MNEs in the clusters. This research is based on case study fieldwork involving semi-structured interviews with relevant actors. In addition, given that the process of automotive clustering involves the interrelatedness of networks among manufacturers and suppliers, this research also discusses the importance of the Japanese *keiretsu* in developing the Indonesian automotive cluster.

Research background and aims

In response to the Asian financial crisis in the late 1990s, the Indonesian government, through its industrial policy, has planned an initiative called *Indonesia Bangun Industri 2025* (Indonesian Industrial Growth 2025), as a platform for reforming national policy. This is intended to support the emergence of the hi-tech manufacturing industry as the prime mover for the national economy;

thus, this initiative is intended to alleviate not only industrial difficulties but also national economic difficulties. The automotive industry is one of the chosen sectors (MOTIRI 2005). It can be argued that none of the Indonesian manufacturing industries has received more policy and analytical attention than the automotive sector (Gunawan 2002; Tarmidi 2004; Wee 2005; MOTIRI 2005). It was one of the first industries to be established in Indonesia and triggered industrialisation from the late 1920s. It is one of the largest manufacturing sectors, and it has recorded rapid growth for most of the 40 years since 1967. However, in 1998–2001 the automotive industry experienced the effects of the economic crisis, as well as being the subject of intense policy intervention, and was variously regarded as a 'spearhead' for technological modernisation, and as a vehicle for the transfer of auto know-how, and production systems and shopfloor management (principally Japanese) knowledge (MOTIRI 2005; Tarmidi 2004).

More recently, Indonesia has missed out on the opportunity to become Southeast Asia's leading automotive nation (MOTIRI 2005). That position is now occupied by Thailand, which in the early 1990s was able to switch more quickly from import substitution to export orientation, and to attract the automotive MNEs in anticipation of a more open Association of Southeast Asian Nations (ASEAN) market for the country. Nevertheless, a technological learning process has undoubtedly taken place in Indonesia behind high protective barriers. As a result, Indonesia has become more efficient in the production of commercial vehicles (notably the Toyota Kijang) and motorcycles compared to the early years of the industry, and a larger market and reduced model proliferation have resulted in higher productivity. The range and quality of components has increased immeasurably, and some component firms are efficient and export oriented, a trend that is likely to accelerate once the current global economic downturn comes to an end.

Research questions

Looking at preliminary research and secondary data, there is a complexity between actors and the process of knowledge transfer in the automotive cluster, which influences the development of the automotive industry in Indonesia. This is due to the nature of the automotive industry specifically in the Japanese global-local integrated production system by *keiretsu*. However, to what extent those actors could provide a significant contribution to the Indonesian automotive cluster and how the interplay between them in the process of knowledge transfer comes about still remains unclear.

In order to understand the context, the research questions were based on the conceptual framework shown in Figure 1.1.

This research aimed to study the automotive industrial cluster in Java in a qualitative way by investigating the involvement of MNEs, through knowledge transfer between subsidiary transplants and suppliers, in accelerating the development of the Indonesian automotive cluster. In this study, Japanese MNEs

4 *Clusters and knowledge transfer*

were chosen as models of exporter industrial organisations because the Japanese automotive industry still has considerable competitive strengths, despite financial difficulties and alliances with foreign partners. Additionally, the issues of technology transfer, skill transfer, and national policies in industrial development are attracting more widespread interest and the automotive industry remains a major global industry.

Indonesia, as part of the major Japanese automotive productive network based in the Southeast Asian region, has been chosen to offer a discussion about the flows of knowledge and technology in the automotive cluster where the Japanese firms have been dominant. Moreover, while numerous studies have been published about the influence of the Japanese automotive industry in other countries, few have appeared on the influence of the Japanese automotive firms in Indonesia.

Using this conceptual framework, I intend to understand the process and development of knowledge transfer in the Indonesian automotive cluster. Since

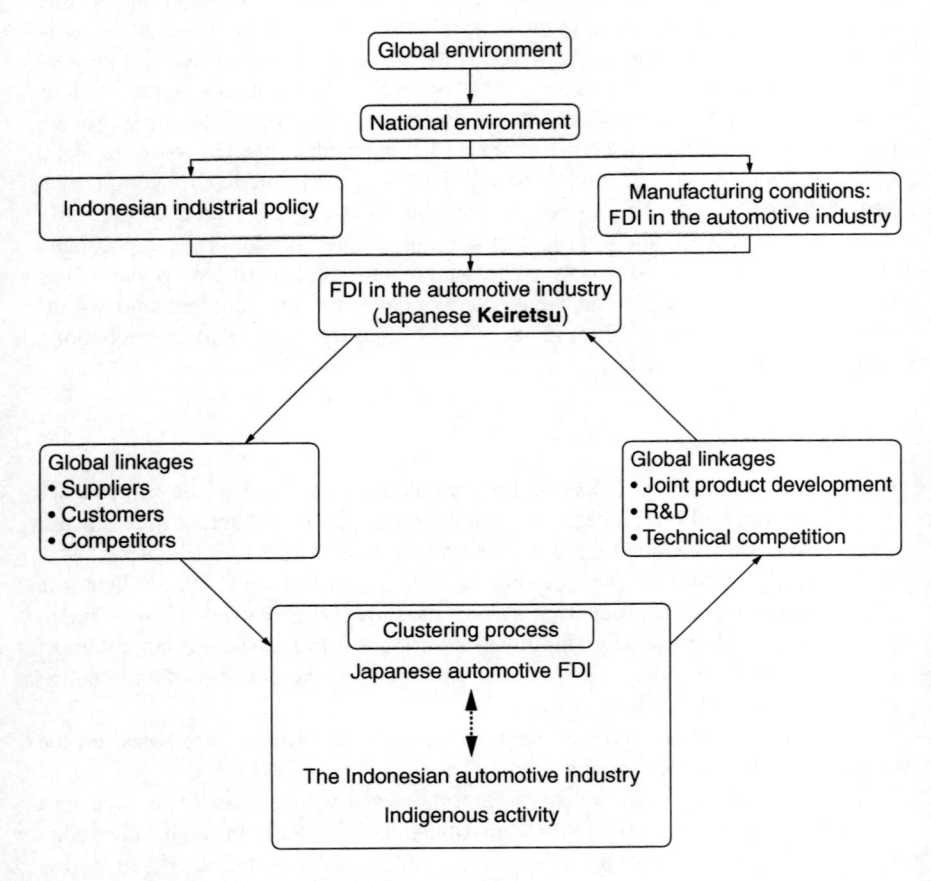

Figure 1.1 Conceptual framework before fieldwork.

the development of a cluster cannot be understood without a thorough insight into the characteristics of the region as a whole, it is therefore important to draw up a frame of analysis to consider several aspects and study their interrelations.

To investigate what is happening in Indonesian automotive cluster (i.e. the process of knowledge transfer) and in what way the global-local production system accelerates the Indonesian automotive cluster, the research questions are as follows:

1 **Context:** How has the global-local network in the automotive industry impacted on the automotive cluster in Java?
2 **Process:** How have the key actors (Toyota and Honda) been involved in fostering technological change and knowledge-technology transfer in the Indonesian automotive cluster (through localisation projects and suppliers)?
3 **Outcomes:** Using two case studies of Toyota and Honda as MNEs in the automotive industry, how have they contributed to the development of the Indonesian automotive cluster (new knowledge, new skills)?
4 **Recommendation:** How can the cluster policy be improved and what lessons can be learned?

Book structure

This chapter has detailed the research background and aims, and listed the research questions. Chapter 2 provides a theoretical framework for the cluster theory literature. Chapter 3 provides the literature on Japanese automotive industry and its knowledge transfer. From the literature base in Chapter 2, the cluster strategy is discussed and subsequently used in Chapter 4, Chapter 6, Chapter 7 and Chapter 8 to analyse the development, implementation, and outcomes of the Indonesian automotive cluster. Additionally, the literature on knowledge transfer and the Japanese automotive sector in Chapter 3 is used in Chapter 5, Chapter 7, Chapter 8 and Chapter 9 to analyse the processes of Japanese MNEs in the automotive industry and the outcomes of the two case studies.

Chapter 4 outlines the literature on the Indonesian context mainly covering the Indonesian macroeconomy and industry along with cluster strategy as an important agenda in Indonesian government. This chapter is also linked with Chapter 6 whereby the discussion of industrial agglomeration in the Java region is elaborated alongside the role of industrial estates as the foundation for developing the cluster strategy.

Chapter 5 provides methodology based on the research questions that have arisen from literatures reviewed in Chapter 2 and Chapter 3 and breaks them down into their constituent parts. It also outlines how the research came about and discusses the methodological approach and research design, before looking at the methods utilised to answer the research questions. The chapter concludes by outlining the limits of the study and the nature of the results that are set out in the following chapters. This chapter will be wrapped up with reflections on the work undertaken, including the research process and limitations.

6 Clusters and knowledge transfer

Industrial agglomeration and clusters in the Java region are the theme of Chapter 6. This chapter draws strongly on the literature outlined in Chapter 2 and Chapter 4. Chapter 6 is an extended explanation of cluster strategy in the context of manufacturing locations in the Java region as a prelude to the automotive cluster discussion in Chapter 7 and Chapter 8.

Chapter 7 and Chapter 8 detail case studies of the automotive cluster in the Java region. Chapter 7 analyses the case of four-wheeled vehicles at Toyota, along with the company's networks for developing technology and knowledge through the manufacturer–supplier relationship. Also, this chapter explores the interrelatedness of government–industry–university in enhancing cluster growth in Java. Chapter 8 provides a case study for two-wheeled vehicles at Honda. It explores the technology and knowledge transfer between Honda's local-global network and its supplier in Indonesia. Furthermore, this chapter also highlights the importance of linkages between Honda, government, and university.

Chapter 9 provides a thorough discussion and analysis based on the two case studies in a more global perspective. It links to Chapter 3 and the literature on the Japanese business network. The chapter examines the Japanese global production network and implications for the Indonesian automotive cluster. This chapter also explains the importance of fostering global-local alliances for the Indonesian automotive cluster in order to enhance innovation, learning and network. Furthermore, the explanation of the global context to finding out how important the Indonesian cluster is in the global production network is also explained in Chapter 9.

In conclusion, Chapter 10 draws together the experiences of the two case studies in the automotive cluster and their varying impacts. This chapter then considers the implications of this research in theoretical terms, through investigating the car and motorcycle clusters and the application of cluster strategy, and the usefulness and 'fit' of different concepts employed to explain clusters and clusters policy. Also, this chapter provides an argument on how to make sense of the global integrated production system from a host country perspective, and outlines areas for future research.

2 Theoretical framework on clusters

This chapter begins by examining the concept of the cluster, its definition and the relation of clusters to the other relevant concepts such as competitive advantage, networks, and the importance of proximity and boundaries. Chapter 2 also underlines the classification of clusters, the benefits from clusters, and different cluster analysis methods.

I then go on to highlight the relation between clusters and policymaking. I emphasise how clusters might impact industrial policy, how clusters can contribute to industrial development, and what risk is involved in implementing a regional cluster strategy.

Additionally, I continue to explain the study of cluster-based policy in developing countries and the importance of FDI (foreign direct investment) which currently lack attention in the cluster literature.

In conclusion, I explain the importance of pulling all the above together in order to understand the connection between the theoretical framework on clusters and the research undertaken.

Cluster definition

The term 'cluster' is not unique to economics, but also (and more frequently) appears in statistics or the computer sciences. In its plain and most general meaning, a 'cluster' is simply defined as a 'close group of things' (*Concise Oxford Dictionary* 1982). In economics, the cluster concept usually implies reference to a particular hypothesis which states that the geographic agglomeration of economic activity may lead to improved technological or economic performance of the units engaged (Coe *et al.* 2007; Solimano 2008).

The cluster hypothesis in its economic sense is based on the work of Alfred Marshall (1890), who explained the development of industrial complexes by the existence of positive externalities within agglomerations of interrelated firms and industries. These externalities are caused by three major forces: knowledge spillovers between firms, specialised inputs and services from supporting industries, and a geographically pooled labour market (Krugman 1991a, 1991b; Krugman and Venables 1995; Fujita *et al.* 1999; Fujita and Krugman 2004).

8 Theoretical framework

Marshall (1890) is commonly cited as the first to mention the occurrence of spatially concentrated industries. The influential works by Marshall (1890) and Weber (1929) on agglomeration economies and more recent contributions by Porter (1990, 1998a, 1998b, 1999c, 2000) on clusters have underpinned cluster studies in the search for explanations of the competitiveness of firms, regions and nations.

According to Marshall (1890), it is possible to derive a *trinity of externalities* causing competitiveness which are all assumed to increase the performance of the involved firms. Despite a widespread consensus on the benefits of agglomerations, there is a great deal of ambiguity surrounding both the nature and the spatial scale of these externalities. It is argued that only a minority of all externalities are actually localised (Scott 1982; Glaeser 2000; Lall *et al.* 2004; Asheim and Coenen 2006).

The basic cause of agglomeration according to Marshall (1890) is the presence of increasing returns to scale which are external to the firm. Based on this concept there are three principal factors for the formation of clusters: labour market effects, input–output interdependency, and knowledge spillovers.

Accordingly, Marshall's arguments for the presence of industrial clustering have been further developed in the new growth and trade theories, where human capital (and therefore knowledge) is an important part of the production function in an industrial cluster (Bresnahan *et al.* 2001; Klepper 2001; Feldman *et al.* 2005b; Asheim and Coenen 2006). Using modern terminology, Marshall's cluster hypothesis (1890) states that the existence of dynamic complementarities within a system of interdependent economic entities influences specialisation patterns in production: for the reason given above, innovation and growth in one of the economic units can create positive impulses for innovation and growth in other parts of the system as well.

Furthermore, Marshall's contribution in 1890 is a cornerstone of cluster – and especially industrial district – theorising, emphasising the dynamics of external economies associated with learning, innovation, and increased specialisation, differing in this respect from industrial location theory (Weber 1929; Hotelling 1929; Hoover 1937). Also, Schumpeter (1934), although never much concerned with the spatial dimensions of economic activity, pointed to a clustering of innovations in time.

A new wave of popularity for clusters was certainly created by Porter (1990, 1998a, 1998b, 1998c, 2000), who takes up many of the earlier arguments and includes clusters as the focus of his famous diamond. Porter revived the cluster debate, although his approach is more a theory of competitiveness. Yet his contribution was taken as a starting point, offering a holistic framework to show how interdependence affects innovation and growth. This combined with notions and concepts such as embeddedness (Grabher 1993; Malmberg and Maskell 2004), social networks (Scott 1991), and untraded interdependencies (Storper 1995; Storper and Venables 2004), and led to a new emphasis on economies of scale and scope in a geographical context (Krugman 1991b; Pinch and Henry 1999).

Using the intellectual antecedents of industrial districts as a foundation, Porter developed a 'cluster theory' model while seeking to understand corporate strategy issues at a regional level. The theory posits that by grouping firms together, cluster analysis can reveal specialisations of production chains in the local region. Thus, the key premise underlying cluster theory is that, through the exchange of specialised information, increased productivity, innovation and new business formation may be achieved within the regional context (Asheim and Gertler 2005; Pavitt 2005; Caniëls and Romijn 2005).

Furthermore, Porter (2000) also stated that the presence of internationally competitive industries ensures, on the one hand, cost-effective and speedy delivery of components and, on the other hand, horizontal support is given when industries in the region coordinate and share activities and thereby stimulate competition. Thus, regional concentration of industries promotes the flow of information and technological spillovers (Dicken and Malmberg 2001; Lundvall 1988; Martin and Sunley 2006).

Porter (2000) also suggested that economic growth is based on Schumpeterian dynamics where interrelated firms in a geographically limited area become more competitive by constant pressure to innovate and by the benefit of being located close to each other. Therefore, clusters could facilitate innovation and the creation of the conditions necessary for the competitive advantages of an industry to persist (Markusen 1996; Feldman 2005; Feldman *et al.* 2005b; Malmberg and Maskell 2006).

Hence, in current debates, the term 'cluster' is mostly related to this local or regional dimension of networks (Saxenian 2005; Nooteboom 2004; Breschi and Malerba 2005) and is defined in various ways in relation to socio-economic concepts.

Most definitions share the notion of clusters as localised networks of specialised organisations, whose production processes are closely linked through the exchange of goods, services, and knowledge (Cooke 2002; Bathelt *et al.* 2004; Tallman *et al.* 2004; Asheim and Coenen 2006). The informal exchange of information, knowledge and creative ideas is considered to be a particularly important characteristic of such networks (Gertler 2003; Bathelt *et al.* 2004; Saxenian 2005).

Unlike sectors, a cluster unites companies from different levels in the industrial chain (suppliers and customers): for example, in the automotive industry, global-local networking plays a crucial role in coordinating the value chain (Nonaka *et al.* 2000; Klepper 2004; Malmberg and Power 2005), along with support from service units (financial institutions, production-supporting services) and connections with government bodies, semi-public agencies, universities and research institutions.

Clusters and competitive advantage

In the management literature, particularly in economics, attention to geography or location has been minimal (Nicholas and Maitland 2002). Globalisation has created a tendency to regard location as of diminishing importance (Amin and Thrift 1995; Wolf 2004). Nevertheless, more attention needs to be given to the importance of location as a variable affecting the global competitiveness of firms for two reasons (Malmberg *et al.* 1996; Fagerbergh 1998; Malmberg and Maskell 1999b).

First, with the gradual dispersion of created assets, the structure and content of the location portfolio becomes more critical to firms who seek to acquire new assets and simultaneously more efficiently deploy their home-based assets (Fagerbergh 1998; Malmberg *et al.* 1996).

Second, the role of the government becomes more critical as the government needs to understand the growing importance of knowledge-related infrastructure and with it the idea of sub-national spatial units as a nexus of untraded interdependencies (Brown and McNaughton 2002; Leclerc and Meyer 2007; Perdana and Friawan 2007). A government's role is to promote the dynamic comparative advantage of the country's resource capabilities and to work in partnership with firms to improve or replace markets in markets where endemic market failure is most widespread (Malecki and Nijkamp 1988; Benneworth and Charles 2001; Etzkowitz *et al.* 2007.)

There are criticisms of this theory. While Porter describes clusters as the new economics of competition, some scholars argue that what the new geographical economists are saying is not new, and it is most certainly not geography (Jaffe *et al.* 1993; Martin 1999; Malmberg and Maskell 1999b). They claim that the current economic emphasis on increasing returns and spatial agglomeration has been already done (and discarded) by geographers who were busy analysing industrial location back in the 1960s and 1970s.

Along with the agglomeration of industry to advance the competitiveness of the region, Martin (1999) opines that the convergence models examine only one aspect of convergence. Thus, there have been few attempts to unravel the qualitative side of knowledge flows, technology spillovers and migration levels (Jacobs and de Man 1996; Saxenian 2005, 2008). Limited attention is paid to the forces that influence the geographical distribution of industry and economic activity (such as local infrastructure, local institutions, state spending and intervention and disinvestment). In this sense the limitations of the studies are that the architecture of economic agglomeration is missing (Krugman 1991a, 1991b; Lagendijk and Charles 1997). Therefore, it is important to present an overview of the profile of such a regional agglomeration based on empirical evidence (Malmberg and Maskell 1999a, 1999b, Yeung 2003).

Returning to Porter's model (2000) of the effect of location on competitive advantage, he portrays four interrelated influences, depicted graphically as a diamond structure; this diamond metaphor has become common in describing the theory. Porter argues that a parallel improvement in the sophistication of company operations and strategies and in the quality of the diamond provides the microeconomic foundations of economic developments.

A few elements of this framework deserve highlighting to understand the role of cluster in competition (Porter 2000). Factor inputs range from tangible assets such as physical infrastructure to information, the legal system, and university research institute, which all companies draw from. To increase productivity, factor inputs must improve in efficiency, quality, and specialisation to particular cluster areas. Specialised factors, particularly those integral to innovation and upgrading (e.g. research universities), not only are necessary to get high levels of productivity but also tend to be less tradable or available elsewhere (Porter 1998c).

The context from firm strategy and rivalry refers to the rules, incentives, and norms governing the type and intensity of local rivalry (Porter 1998a, 1998b, 1998c). Economies with low productivity are characterised by little local rivalry. Most competition comes from imports; hence, local rivalry involves imitation. Price is the sole competitive variable, and firms hold down wages to compete in local and foreign markets. Here, competition involves minimal investment.

Moving to an advanced economy requires rigorous local rivalry to develop. Rivalry must shift from low wages to low total cost and it implies upgrading the efficiency of manufacturing and service delivery. Competition must shift from imitation to innovation and from low investment to high investment in not only physical assets but also intangibles (skills, technologies). Hence, clusters will play an integral role in these transitions (Gertler 2003; Caniëls and Romijn 2005).

The character of rivalry in a location is strongly influenced by many aspects of the local business environment. Yet the investment climate and policies towards competition set the context. Things such as macroeconomics and political stability, the tax system, labour market policies affecting the incentives for workforce development, and intellectual property rules and their enforcement, contribute to the willingness of the company to invest in upgrading capital equipment, skills, and technology (Solimano 2008). Antitrust policy, government ownership and licensing rules, policies towards trade, foreign investment, and corruption have vital roles in setting the intensity of local rivalry.

In terms of demand, local demand can also reveal segments to the market where firms can differentiate themselves. In the global economy, the quality of local demand matters far more than its size (Krugman 1991a; Krugman and Venables 1995). Clusters of linked industries play a central role in giving rise to demand-side advantages. Hence, the cluster is the manifestation of the diamond at work (Porter 2000). Clusters affect competition in three broad ways that both reflect and amplify the parts of the diamond: (a) increasing the current (static) productivity of constituent firms or industries; (b) increasing the capacity of the cluster participant for innovation and productivity growth; and (c) stimulating new business formation to support innovation and expand the cluster. Many cluster advantages rest on external economies or spillovers across firms, industries, and institution of various sorts; hence, a cluster is a system of interconnected firms and institutions whose whole is more than the sum of its parts.

Moreover, the three broad influences of clusters on competition depend on personal relationships, face-to-face communication, and networks of individuals and institutions that interact (Saxenian 1996; Brown and McNaughton 2002;

Prahalad and Ramaswamy 2004). However, the level of development for making such relationships is far from automatic as formal and informal organising mechanisms and cultural norms often play a role in the functioning and development of clusters (OECD 1999b).

It should be clear that clusters represent a combination of competition and cooperation (Gallaud and Torre 2004; Steiner and Hartmann 2006). Because of the presence of multiple rivals and strong incentives, the intensity of competition within clusters is often accentuated. Yet, cooperation must occur in a variety of areas, according to Porter (2000). Competition and cooperation can coexist because they are on different dimensions or cooperation at some levels is part of winning the competition at other levels.

Clusters offer obvious transaction cost advantages over other forms of organisation. Repeated interaction and informality of contracts within the cluster will foster trust and open communication, while reducing the cost of severing and recombining market relationships (McCann 1998; Maher 2001). Thus, the connection between clusters and competition carries important implications for the economic geography of cities, nations, and groups of neighbouring countries (Porter 1998a; Fujita and Krugman 2004). Internal trade within nations is a powerful force for improving productivity, as is trade with immediately neighbouring countries (Krugman and Venables 1995; Porter 1998b). Therefore, the formation of clusters is a significant element of economic development (OECD 1999a, 1999b; Porter 2000; Malmberg and Maskell 2004).

Clusters and networks

In recent work on networks and industry studies, networks have been recognised as a very important ordering principle in the knowledge-based economy where engagement in networks has several well-documented advantages (Yeung 2000; Kogut 2000; Feldman 2005). Firms and organisations more actively engage in networks as a means to survive in a volatile international market and to compete along with rapid technological change (Van Den Berg et al. 2001a, 2001b; Nooteboom 2004; Saxenian 2005).

In order to benefit from chance opportunities, a firm has to be able to react fast and to engage in partnerships with complementary strengths and capabilities (Gordon and McCann 2000). This is particularly true for firms which are located in clusters of densely networked firms (Malmberg and Maskell 2004), where inter-firm and interorganisational cooperation in networks have different spatial dimensions. Networks can extend worldwide and many network relations between actors can be located in a specific area, region or city (Kim and Von Tunzelmann 1998; Gordon and McCann 2000; Van Den Berg et al. 2001a, 2001b).

Networks are significant regarding innovation as strong international competition and fast technological development urge firms to innovate constantly in terms of integrated production systems across the global network, such as in the automotive industry (Henderson 2002; Klepper 2002, 2004; Pries and Schweer

2004). Participation in a network enables a firm to concentrate on core capabilities and provides access to resources (specific know-how, technology, financial assets, products, market) in other firms and organisations. Thus, networks will help companies to improve their competitive position (Bathelt et al. 2004; Tallman et al. 2004; Saxenian 2008).

In addition, Tichy (1987: 94) argues that clusters consist of nodes linked by a network. These nodes are firms, industries, and other public and private institutions which are linked through supplier–buyer (input–output) relations and various other forms of cooperation. Different types of networks or clusters such as hub-and-spoke or vertically disintegrated structures can be identified, depending on the underlying hierarchical structure. Thus, synergies and spillovers are potential benefits associated with the linkages between the various actors or networks and the benefits from geographical proximity (agglomeration economies) (Cooke et al. 2003; DuPuy and Torre 2006). These synergies and spillovers, which are external to the firm but internal to the cluster, serve as incentives for joining an existing cluster or taking part in its initiation (Breschi and Lissoni 2001; Gallaud and Torre 2004; Steiner and Hartmann 2006). Examples of externalities with particular relevance to the automotive sector are skilled labour markets or specialised business services (Larsson 1999; West 2000; Pries and Schweer 2004).

Subsequently, the cluster provides a constructive and efficient opportunity for discussion among related companies, their supplier, government, and other institutions (Goyal and Joshi 2003; Nooteboom 2004). Because of externalities, public and private investments to improve cluster circumstances benefit many firms. Seeing a group of companies and institutions as a cluster also highlights opportunities for coordination and mutual improvement in areas of common concerns with less of a risk of distorting competition and limiting the intensity of rivalry (Tichy 1998; Ernst and Ravenhill 2000).

Thinking about the influence of location on competition has been based on relatively simple views of how companies compete (Feldman 2000; Dicken 2003; Lall et al. 2004; Prahalad and Ramaswamy 2004). Competition is seen as largely static, resting on cost minimisation in relatively closed economies. Here the comparative advantage in factors of production is decisive, which means that increasing returns to scale play a central role.

Close linkages with buyers, suppliers, and other institutions are essential, not only to efficiency but also to the rate of improvement and innovation. Location affects competitive advantage through its influence on productivity, particularly productivity growth (Scott 1982; Von Hippel 1998; Feldman et al. 2005a; Asheim and Gertler 2005).

Nevertheless, the importance of location for companies to compete is strongly influenced by the quality of the microeconomic business environment. Some aspects of the business environment (e.g. infrastructure, the legal system, corporate tax rates) cut across all industries (Lundvall and Christensen 1999). These economy-wide (or horizontal) factors are important and represent the binding constraints on competitiveness in developing economies. In more advanced economies and increasingly elsewhere, however, the more decisive aspects of the business environment for competitiveness are often cluster specific (e.g. the presence of particular types of suppliers, skills, or university departments) (Asheim and Isaksen 2002).

Proximity and industrial clusters: the impact on networks and technology

The importance of proximity for institutional support for the networks will play a major role in regional economic and technological development (Ahuja 2000; Nooteboom 2004; Powell and Grodal 2005). In the economic literature this idea gives birth to an increasing use of concepts such as local networks, localised systems of production and local systems of innovation (Castells 1996; Cooke 1995; Ernst 2000; Nooteboom 2003; Gertler and Wolfe 2006).

All these concepts rest on the importance of geographical proximity for actor relations in a network setting (Boschma 2005; Audretsch and Lehmann 2006), as they are convinced of the importance of these local networks (Cooke 2004; Belussi 2006; Steiner 2006). Accordingly, local networks can be divided into two categories:

1   *Spontaneous local networks*
    These are groupings of local actors around one or several joint economic projects, according to a non-market form of organisation. The links are generally not based on contracts or complete explicit agreements, but rather on processes of cooperation or collective learning. Their main purpose is a common interest in the production of a good, in sharing a technique, or in the search for information needed by all the members. The exchange mainly relates to the transfer or sharing of knowledge and is effected through trust relations.
2   *Institutional local networks*
    These correspond to structures settled by public bodies in order to support the firms. They concern flexible forms of organisation, founded on a common acceptance of the rules that engage the participants, among which one can make a distinction between the producers and then users of information, and technological knowledge. The link between the participants in the network is materialised by an adherence to, as well as utilisation of, the services offered by an organising cell which also plays a part in animating the whole network. There are general and specialist networks.

Subsequently, there is a difference in technological policy according to whether or not spontaneous local networks already exist. When they do not exist or are poorly developed, policy aims to promote them, even to create them, by means of incentives or voluntarist policies (Yeung 2000; Breschi and Malerba 2005). When they already exist, the objective is to support their development by, in particular, supporting transverse cooperation between partners belonging to different worlds (industry, research, universities, technical centres) (Yeung 2000; Ahuja 2000; Breschi and Malerba 2005). In both cases, the objective is to connect a spontaneous network of economic actors with an institutional environment promoted by the local authorities (Jarillo 1993; Breschi and Malerba 2005).

Nevertheless, Torre and Rallet (2005) argue that the uneasy installation of local networks of innovation supported by public policies is difficult to promote, and outline two main obstacles.

The first difficulty is to set up transverse cooperation between local actors of various kinds (entrepreneurs, researchers, trainers). Tacit knowledge is more easily transmissible within a professional world (even at distance) than between different worlds (even in proximity) (Nonaka and Takeuchi 1995; Nonaka et al. 2000; Gertler 2003; Cooke 2006). Consequently, the diffusion of knowledge and technology assumes that an organisational proximity exists between the actors, previous relations founded on professional links, irrespective of local content.

Thus organisational proximity does not necessarily have local foundations (Nooteboom 2003; Malmberg and Power 2005). For reasons linked with the way in which the local systems were constituted, actors often engage in cooperation with partners external to the region. They are accustomed to cooperate, a practice which results in mutual knowledge of people and organisations, and they have common work procedures which have proved reliable. Putting actors who are geographically close to each other in contact is not sufficient if they did not have organisational relations previously; thus, the history of local relations counts (proximity matters) as well as the history of the non-local relations (distance matters) (Boschma 2005; DuPuy and Torre 2006).

Additionally, in terms of networks, Porter (2000) describes how clusters of densely networked firms serve global markets while deriving their strength from a regional basis. He discerns four conditions as essential in that development: factor conditions as described in his diamond model. In particular, the interplay of competition and cooperation is fundamental. Excessive competition may be destructive but the same holds for excessive cooperation when it degenerates into the formation of cartels (Harrison 1994; Cooke 1995; Van Den Berg et al. 2001a, 2001b).

Subsequently, other scholars argue that in clusters, another role is played by non-inter-firm linkages: links with government-supported scientific institutes, ties with the scientific community, and professional associations are important factors in a cluster's performance (Lazonick 1992; Boekholt 1994; Gertler 1995; Gallaud and Torre 2004).

Yet the question remains why proximity still seems to matter in networks, where modern communication technology theoretically permits spatial dispersion (Gertler 1995; Feldman et al. 2005a; Boschma 2005; Dupuy and Torre 2006). It is because face-to-face contacts are still very important as sources of technological information and in the exchange of tacit knowledge (Leonard 1984; Malmberg et al. 1996; Nonaka et al. 2000; Gertler 2003). Spatial proximity greatly enhances the possibility of such contacts. Second, cooperation between actors requires mutual trust. This holds true particularly when sensitive and valuable information is exchanged, for instance in product development of a new design in the car industry or in a joint innovation project (Tassey 1991; Pries and Schweer 2004).

Cluster boundaries

Clusters can occur in many types of industries, in smaller fields, even in some local industries, such as restaurants, car dealers, or antique shops (Benneworth and Henry 2003; Tallman *et al.* 2004; Breschi and Malerba 2005). They are present in large and small economies, in rural and urban areas, and at several geographic levels (e.g. nations, states, metropolitan, regions, and cities). Moreover, clusters can also take place in both advanced and developing economies, although clusters in advanced economies tend to be far more developed (Ernst 2000; Parrilli 2006; Metcalfe and Ramlogan 2006).

Cluster boundaries rarely conform to the standard industrial classification system, which fail to capture many important actors in linkages and across industries (Tichy 1998). Because parts of clusters are often put into different traditional industrial or service categories, significant clusters might be obscured or even unrecognised. Consequently, the appropriate definition of a cluster can differ in different locations, depending on the segment in which the member companies compete and the strategies they employ (Schmitz 2000; Van Den Berg *et al.* 2001a; Martin and Sunley 2003). Clusters can also be examined in various fields of aggregation (e.g. agriculture clusters, wine clusters), thereby exposing different issues (Lagendijk and Charles 1997; Hoen 2002; Giuliani *et al.* 2005; Giuliani 2006).

The boundaries of clusters continually evolve as new firms and industries emerge, established industries shrink or decline and local institutions develop and change (Malmberg 2003). Technological and market developments give rise to new industries, create new linkages, or alter served markets. In addition, the importance of viewing economies through a cluster lens instead of more traditional groupings such as companies, industries, SIC codes, and sectors (e.g. manufacturing, services) is because the cluster as a unit of analysis is better aligned with the nature of competition and the appropriate roles of government (Hoen 2002; Benneworth and Charles 2001; Coe *et al.* 2007; Kuznetsov and Sabel 2007). Clusters, broader than traditional industry categorisations, capture important linkages, complementarities, and spillovers in terms of technology, skills, information, marketing, and customer needs that cut across firms and industries. These externalities create a possible rationale for collective action and a role for government (Moulaert and Swyngedouw 1991; Henderson *et al.* 2001; Rikard and Urban 2008).

Such connections across firms and industries are fundamental to competition, to productivity, and particularly to the direction and pace of new business formation and innovation. Most cluster participants are not direct competitors but rather serve different segments of industries. Yet they share many common needs, opportunities, constraints, and obstacles to productivity (Steiner and Hartmann 2006).

Theoretical framework 17

The cluster provides a constructive and efficient space for discussion among related companies, their suppliers, government, and other institutions (Goyal and Joshi 2003; Nooteboom 2004). Public and private investments to improve cluster circumstances benefit multiple firms. Seeing a group of companies and institutions as a cluster also highlights opportunities for coordination and mutual improvement in areas of common concerns with less of a risk of distorting competition and limiting the intensity of rivalry (Tichy 1998; Ernst and Ravenhill 2000).

Benefits from clusters

Today, the nature of agglomeration economies has shifted towards the cluster level and away from either narrow industries or urban areas per se (Porter 2000; Storper and Venables 2004). Thus, many treatments of agglomeration economies rest on cost minimisation due to proximity to inputs or proximity to markets (Boschma 2005; Feldman 2005). Detailed cluster benefits are as follows:

1 Access to specialised inputs and employees

Clusters can provide superior or lower cost access to specialised inputs such as components, machinery, business services, and personnel compared to vertical integration, formal alliances, or importing inputs from distant locations (Porter 1998a). Here, the cluster will have a spatial role that can be an inherently more efficient or effective means of assembling inputs than the alternatives. Thus, given the inherent benefits of clusters, forces encouraging local supplier development are strong and constituent firms have an incentive to encourage entry of new suppliers through local investment by distant suppliers. More importantly, however, the presence of a cluster not only increases the demand for specialised inputs but also increases the supply. Thus, the availability of specialised personnel, services, and the number of entities creating them often is far greater within a cluster than elsewhere despite the greater competition (Van Den Berg et al. 2001b; Solimano 2008; Saxenian 2008).

2 Access to information

Proximity, supply and technological linkages, and the existence of repeated personal relationships and community ties fostering trust, facilitate the information flow within a cluster (Gallaud and Torre 2004; Dupuy and Torre 2006). Furthermore, clusters can accommodate extensive market, technical, and other specialised information accumulation in firms and local institutions (Boschma 2005). Thus, obtaining information about current buyer needs and sharing information with other cluster participants are benefits of clusters.

3 Complementarities

Clusters increase productivity not only through the acquisition and assembly of input but also through facilitating complementarities between the activities of cluster participants (Hoen 2002; Maskell *et al.* 2006). Some of the most important examples are complementary products for the buyer, where the location of firms and industries within a cluster makes it easier to achieve product-service coordination and creates internal pressures for improvement among parts of a cluster; and marketing complementarities, where a group of related firms or industries can create joint marketing (e.g. trade fairs, firm referrals, marketing delegation), which can enhance the reputation in a particular field and makes the buyer more aware of the vendor or manufacturer based there (Gertler and Wolfe 2006; Malmberg and Maskell 2006; Poon *et al.* 2006). The presence of multiple sources for a product or service in a location can reduce perceived buying risks by offering buyers the potential to multi-source or switch vendors. Another example is complementarities due to a better alignment of activities among cluster participants, which means providing easier linkages with suppliers, channels, and downstream industries within a cluster than among dispersed participants (Dahl and Pedersen 2003; Solimano 2008; Saxenian 2008).

4 Access to institutions and public goods

Firms within a cluster can access specialised infrastructure, or advice, from experts in local institutions at very low cost. Indeed, the information built up in a cluster can be seen as a quasi-public good as accessing it involves some costs although well below full cost. Additional quasi-public goods in clusters are often the result of private investment in training programmes, private infrastructure, quality centres, etc. Here, private investments in cluster-specific public goods or quasi-public goods are common because of the collective benefits perceived by cluster participants. Often, such private investments in public goods take place through trade associations or other collective mechanisms (Steiner and Hartmann 2006).

5 Incentives and performance measurement

Clusters improve incentives within companies for several reasons. The first is competitive pressure, when the pride and desire to perform well in a local community motivates firms to compete with each other. Clusters also make it easier to measure the performance of in-house activities because there are local firms that perform similar functions. Thus, managers can compare internal costs, as well as lower employee performance monitoring costs, to those of others locally. Therefore, the accumulation of knowledge for financial monitoring will be improved. In addition, due to repeated interaction, easy spread of information and reputation, and desire for standing in the local community, interaction among cluster participants is more likely to be constructive and reflect long-term interests (Nonaka *et al.* 2000; Malmberg and Power 2005).

More important than the benefits of clusters in current productivity is the role of clusters in innovation and productivity growth (Lundvall 1998; Asheim and Isaksen 2002; Feldman 2005). According to Porter (2000), cluster participation offers many potential advantages in innovation and upgrading compared to an isolated location, although it involves some risks as well. Furthermore, the advantages of a cluster for new business formation can play a major role in speeding up the innovation process (DeBresson and Hu 1999; Ernst and Kim 2002; Saxenian 2002). Large companies often face various constraints and impediments to innovating. Spin-off companies often pick up the slack, sometimes with the blessing of the former companies. It is common to see larger companies in a cluster develop close relationships with innovative smaller ones, help establish them, and even acquire them if they become successful.

Another advantage for innovation is the competitive pressure, peer pressure, and constant comparison that occur in geographically concentrated clusters (Feldman 2000; Malerba 2002; Boschma 2005). The similarity of basic circumstances (e.g. labour costs, utility costs), combined with the presence of multiple rivals, forces firms to seek creative ways in which to differentiate themselves. Here, pressure to innovate is elevated. Individual firms in the cluster have difficulty in staying ahead for long, but many firms are often able to progress much faster than those based at other locations (Asheim and Isaksen 2002; Cooke *et al.* 2003; Cooke 2004).

Furthermore, Porter (2000) explains that firms within a cluster are able to more clearly and rapidly perceive new buyer needs. Thus, firms in a cluster can benefit from the concentration of firms with buyer knowledge and relationships, the juxtaposition of firms in related industries, the concentration of specialised information-generating entities, and buyer sophistication. Cluster firms can often discern buyer trends faster than isolated competitors. Silicon Valley and Texas-based computer companies, for example, are quick to plug into customer needs and trends (Saxenian 1996; Porter 2000; Bresnahan *et al.* 2001; Saxenian 2005).

In addition, cluster participants also offer advantages in perceiving new technological, operating or delivery possibilities (Saxenian 2005, 2008). Participants can be exposed to richer insights into evolving technology, component and machinery availability, and service and marketing concepts. Ongoing relationships with other entities within a cluster (including universities) foster such learning. Therefore, direct observation of other firms is facilitated. By contrast, the isolated firm faces higher costs and steeper impediments to assembling insight, as well as a greater need to create knowledge in-house (Audretsch and Feldman 1996b; Harrison *et al.* 1996; Jaffe *et al.* 1993).

Firms within a cluster can experiment at lower cost or delay large commitments until there is greater assurance that a new product, process, or service will pan out (Sugiyama and Fujimoto 2000; Klepper 2004). By contrast, a firm relying on distant outsourcing faces greater challenges of contracting, securing delivery, obtaining associated technical and service support and coordinating across complementary entities (West 2000; Pries and Schweer 2004).

Accordingly, many new businesses are formed in existing clusters rather than in isolated locations (Porter 2000; Tallman *et al.* 2004). This occurs for a variety of reasons. The inducement to entry is greater within the cluster because there is better information about opportunities. The existence of a cluster signals an opportunity. Individuals working somewhere in or near the cluster more easily perceive new gaps to fill in products, services, or suppliers (Power and Lundmark 2004; Rikard and Urban 2008). Having had the benefit of these insights, these individuals more readily leave established firms to start new ones aimed at filling the perceived gaps.

In conjunction with the opportunities perceived in the cluster, necessary assets, skills, inputs and staff are readily available at the cluster location and are assembled more easily there. Hence, the barriers to entry are lower than elsewhere. Lower entry barriers, the existence of multiple potential local customers, established relationships, and the presence of other local firms can reduce the perceived risk of entry. On the other hand, the barriers to exit the cluster can also be lower due to a lower need for specialised investment, a deeper market for specialised assets, and other factors (Drejer et al. 1999).

Nevertheless, cluster participation can also delay innovation (Morgan 1997a, 1997b; Malmberg and Power 2005). It happens when a cluster shares a uniform approach to competing: 'groupthink' can reinforce old behaviours, suppress new ideas, and create rigidities that prevent adoption of improvements (Glasmeier1991; Breschi and Lissoni 2001). Moreover, clusters might not support truly radical innovation, which tends to invalidate the existing pools of talent, information, suppliers, and infrastructure (Cooke 2001; Benneworth and Henry 2003). In these circumstances, a cluster participant might be no worse off, in principle, than an isolated firm, but the firm in an established cluster might suffer greater barriers to perceiving the need to change and inertia against severing past relationships that no longer contribute to competitive advantage (Porter 2000; Metcalfe and Ramlogan 2006).

Cluster analysis methods

The cluster approach offers an alternative to the traditional sectoral approach; it offers another way of looking at the economy and at innovations (Gammeltoft and Aminullah 2004; Malerba 2005; Asheim and Coenen 2006). While the sectoral approach focuses on horizontal relations and competitive interdependence (relations between direct competitors with similar activities and operating in the same product markets), the cluster approach also focuses on the importance of vertical relationships between dissimilar firms and symbiotic interdependence based on synergism (Hatch and Yamamura 1996; Ernst and Ravenhill 2000).

Roelandt et al. (1999) distinguish two empirical methods for identifying clusters: the *monographic method* and the *input–output method*. DeBresson and Hu (1999) add a third method, the graph method. Since the innovation data required for this method are not available, the method for this study has to be one of the other two methods.

The monographic method generally uses a cluster chart based on Porter's diamond to identify the most important clusters in an economic system. It can be used to identify innovative clusters as well as value-added chains. This method generally involves interviews, surveys, and case studies; the techniques used are more qualitative than quantitative (Kiba and Kodama 1991; Steiner and Hartmann 2006).

By contrast, the input–output method is a more quantitative method. Based on the linkages in an input–output table, the sectors that use each other's products are identified and grouped into clusters. This method can be easily applied for several years or countries, which adds to the objective character of the method. Besides this advantage, there are other positive features associated with the input–output method. Input–output tables are available for most countries, with data thus more easily available than those required for the monographic method. Furthermore, the input–output method is generally a straightforward method that is easy to apply.

Nevertheless, by specifying strict boundaries for industries (mostly based on some statistical convention), research fails to take into account the importance of interconnections and knowledge flows within a network of production (Saxenian 2002; Ernst and Kim 2002; Van Haken 2005; Bryson and Henry 2005).

Microstudies begin with similar broad insights into why firms successfully colocate and then redirect their principal focus to how groups of similar sector firms cooperatively share production capacities, markets, labour and technologies, reserving for such Italianate arrangements the term cluster (Malerba and Orsenigo 1990; Morrison and Rabelloti 2005).

Such studies by definition limit attention to evidence of currents flowing among similar sector firms that are best detected up close and at fairly small geographic scales. This approach restricts its view to a single visible collection of similar sector firms; thereby overlooking linkages that some of its members may have with regionally colocated firms from very different sectors, or robust clustering of other sectors.

It is therefore no surprise that microstudies only document one cluster per region (Meeuwsen and Dumont 1997; Porter 1998c). This apparent indifference to the presence of additional clusters, particularly those based on alternative criteria or detectable only from a wider spatial view, is mainly the result of micro-oriented investigations of a priori cluster definitions (Burgess and Venables 2004). This implies that significant instances of region-wide industrial clustering go unrecognised by microstudies (Giuliani et al. 2005).

Microclusters include the highly localised examples of the 'Third Italy', where nearly all economic activity takes place in small, tightly bounded territories – usually within industrial districts (Steiner 1987; Tichy 1998). The truly homogenous nature of an 'Italianate' cluster means that the surrounding territory is wholly devoted to it. Single microclusters consisting of one or a few sectors also extend to more diffused locations. However, microclusters based on industrial district assumptions are less likely to afford an adequate strategic view in the USA and other Western economies, simply because of the more heterogeneous mix of industries that constitute their regions (Saxenian 1996; Steiner 2006; Martin and Sunley 2006). The handful of sectors included in a typical microcluster provides a limited view of the full regional economy and how it functions.

There is no universally defined methodology for conducting microcluster studies. However the very specificity of microstudies also limits their usefulness to apparently similar clusters located in different regions that may have quite distinct institutions and leadership dynamics (Rosenthal and Strange 2003).

Microstudies tend to revolve around the needs of the focal industries to survive or thrive in their settings. Most inquiries are therefore geared to learning what is needed to act decisively in their specific economic and regional environments. These studies are attempts to provide useful specificity and detail, and of how subtle connections are made, networks are maintained, and interpersonal assets are translated into cluster advantages of the utmost importance to sponsoring clients. These interests may be at odds with host regions that wish to restructure their economies away from the most vulnerable towards the most promising clusters.

Even though the industry cluster concept does not constitute a new self-contained model of regional development, it does represent a comprehensive description of how economic and geographic interdependences are integral to regional growth and development processes (Fujita and Krugman 2004).

Macro-based definitions start instead by analysing transaction networks of universally documented flows that channel various interactions among all active industrial members of a trade-based economy, often multinational trading blocs such as NAFTA (North American Free Trade Agreement), or the EU. These networks are to be distinguished from the concern for internal layerings of a national economy that animated much early French *filière* research (Jacobson and Andrèosso-O'Callaghan 1996, pp. 118–120). Network flows consist of communication and transportation links, technology or patent diffusion and exchange, or inter-industry trade. Various analytic techniques are applied to observed flow data as a means of detecting repeated transaction tendencies among subsets of all firms and industries, which then defines a flow-based cluster.

Value chain macrostudies capture highly probable interdependencies, mainly the inter-industry trade of advanced industrial goods that Krugman (1991a, 1991b) and other trade theorists (Fujita and Krugman 1995; Fujita *et al.* 1999) now stress, thus drawing attention to *industrial trade clusters*. Comparative production advantages brought about by reduced trade barriers, increasing returns, declining real costs of transportation, and the revitalised significance of agglomeration economies can both increase regional (or national) focus on what can be produced competitively for export trade, and enable things that cannot be made competitively to be purchased at lower cost.

Regions can become highly significant nodes in expanding trading regimes, particularly as they specialise in the production of a narrower, trade-linked range of products, which enables such production in ever larger volumes, for sale to national or global markets (Andersen and Christensen 2005). Increasing returns permit the growth of larger scale firms and industries in specialised regions to produce more efficiently and cheaply, and to ship easily to open markets (Breziz and Krugman 1993; Amin 2002). Industries dependent on strong supplier ties colocate in the same or nearby regions to take advantage of joint production factors, creating afresh the familiar preconditions for trade-driven industrial clusters.

The 'current of cooperation' observed in microclusters (Roelandt et al. 1999) are instead presented as 'collaborative arrangements' among the majority of industrial trade cluster firms, because even fiercely competitive firms seldom prosper, nor survive very long, in absolute isolation.

Competitive firms are reliant upon supplier agreements to make timely deliveries of high-quality intermediate inputs and services. They also learn about and continually incorporate key technological innovations and product improvements available through collaborative contacts with nearby suppliers of inputs, production machinery, and other forms of capital equipment (Rutherford 2000).

Finally, firms rely upon informal exchange of information, technology, and specialised knowledge that routinely flows between trading partners, a practice widely acknowledged to be an important source of unpriced externalities that can improve individual and collective competitiveness, and that extends even to distant trade partners (DeBresson and Hu 1997; Storper 1997).

The analysis of linkages and interdependence between actors in value chains or innovation systems can be carried out at different levels of analysis and with different techniques, depending on needs and the questions to be answered (Benneworth and Charles 2001; Benneworth and Henry 2003). Some studies focus on the firm level and analyse the competitiveness of a network of suppliers around a core enterprise. This analysis is used to make a strategic analysis of the firm and to identify missing links or strategic partners when innovation projects encompass the whole chain of production (Ernst and Kim 2002; Coe et al. 2007). As well as the level of analysis, cluster methodologies differ in techniques, with several ways to conduct research in cluster studies. Thus, the research techniques could be applied as follows:

1 *Input–output analysis*: focusing on trade linkages between industry groups in the value chains of the economy (Metcalfe 1995; DeBresson 1996; Bresnahan et al. 2001; West 2002).
2 *Graph analysis*: founded on graph theory and identifying cliques and other types of network linkages between firms or industry groups (DeBresson and Hu 1997, 1999; Meeuwsen and Dumont 1997).
3 *Correspondence analysis*: for instance, factor analysis, principal components analysis, multidimensional scaling and canonical correlation (Pedersen et al. 2003; Van Haken 2005).

All these techniques aim at identifying groups or categories of firms or industries with similar innovation styles. Furthermore, the qualitative case study approach such as that used by Porter could be used as a complementary method in conducting research in cluster studies (Porter 2000; Nicholas and Maitland 2002; Van Haken 2005).

In relation to Marshall's model trinity (1890), elaborations are as follows:

1 *Labour-market advantages* result from the fact that the firms of the cluster demand similar qualifications from their employees. Firms profit from the

existence of a labour-market cluster in two ways. First, they need not train employees themselves, but can buy them from a market for skilled labour; such qualifications are high. This allows the cultivation of specialised institutions. Formal and informal contacts between specialists will improve skills further if a critical mass of firms and employees can be reached. Second, firms will profit from technology spillovers, resulting from the transfer of skilled people from other firms. Workers profit from learning effects – higher qualifications result in higher wages – and they profit additionally from thick-market externalities, when the market for their skills is large. To sum up: the labour market advantage of a cluster is the two-way network that defines, creates, and supplies skills and produces thick markets and externalities.

The ability of highly specialised high-skill industries to buy expert skills in an efficient labour market is more important than ever before. Even if it pays to educate and train staff within a firm, additional demands for qualifications can be served only after a time consuming training period, which places heavy constraints on flexibility. The concentration of skills in the cluster areas, therefore, is a huge competitive advantage. The theory of the learning curve and of learning by doing (Lucas 1988) applies even more to clusters than it does to firms.

2 The second advantage of clusters, *input–output interdependence*, enables firms to outsource work (processes) which others can execute more efficiently. The higher efficiency of those suppliers rests on two pillars: economies of scale and economies of specialisation. Both result from the concentration of demand for intermediate products within the cluster. Therefore, a cluster profits from a second network of exchange of intermediate goods and services, relying on an information network as the basis for this complicated but highly efficient division of labour between firms.

Input–output interdependence was highly important, but has been widely lost in an era of specialisation and outsourcing – keywords in both modern macroeconomics and business economics. New growth theory emphasises the (external) economies of scale resulting from the existence of highly specialised suppliers (Romer 1986), and most modern business economics (over) emphasise concentration on core activities, outsourcing, and just-in-time production (Womack *et al.* 1990; Liker and Meier 2005). All this implies that individual firm competitiveness relies heavily on the existence of a surrounding cluster.

3 *Technology spillovers*, the third component of clusters, result from the existence of research and development (R&D) specialists in the region, which can exist because of the concentrated demand for their services. The mobility of this 'third network' of skilled specialists (a technology network) supplements the labour-market network and the input–output network. Any one of these three networks supports and improves the other two.

The technology network has become the most important constituent of clusters today. New growth theory emphasises the importance of research for economic growth (Romer 1986, and the theory of innovation has scrapped the old linear model (Myers and Rosenbloom 1996). Innovation needs an integrated and interactive approach that blends scientific, technological, socio-economic, and even cultural aspects with organisational capabilities in rapidly changing environments. Firm-specific knowledge embedded in its workforce has to be blended with generally accessible knowledge. But this knowledge has to be detected – information may be an almost free good – but not knowledge. This integrated and interactive approach to innovation is facilitated by geographical proximity and frequent face-to-face contacts between people with similar interests, that is, through the existence of a cluster. Clusters can therefore be described as learning organisations (Coombs *et al.* 1996) and this is completely consistent with the fact that new technologies also arise and diffuse in clusters. Important new technologies cannot be applied step by step but only in a big bang; this gives greenfield investments a head start over existing plants and restricts the adaptability of clusters.

There is also a fourth factor for the importance of clusters today, namely the *new organisational principles of firms*. This increasing complexity of products and processes, the compulsion to innovate continuously, and the pressure of financial markets to maximise shareholder value sky-rocketed the complexity of management and forced firms to concentrate on their core activities. To be viable, however, such strategy requires the firm to be embedded in a cluster. Today's clusters are a substitute for yesterday's hierarchical firms, which have proved to be unmanageable. If a cluster works well, firms can profit considerably from the R&D expenditures of other firms (Audretsch and Feldman 1996b).

Clusters are therefore more important than ever for high-tech economies. They allow workers, firms, research institutions, and countries to profit from specialisation, while the large integrated markets allow more division of labour than ever before (Power and Lundmark 2004; Rikard and Urban 2008). Specialisation increases efficiency but it increases risk as well: a specialist is highly vulnerable to shifts in demand or to innovations rendering his or her skills valueless (Tichy 1987, 1998). If this occurs, the cluster ends. To prevent this and to ensure lasting instead of temporary profits for clusters, ways must be found to keep a cluster young and flexible (Hoen 2002).

Industrial clusters, which consist of firms linked actively together and in close spatial proximity, have become a very popular concept, one that is now widely embraced (Gordon and McCann 2000; Feldman *et al.* 2005b). It is difficult to identify another relevant concept that appeals to such a broad spectrum of academic disciplines, professions, and even lay people.

Different perspectives are brought to the cluster concept by the swelling ranks of its many diverse adherents (Maskell 2001; Martin and Sunley 2003; Malmberg and Maskell 2004; Belussi 2006). Certain adherents favour only particular versions of what constitutes the essence of a 'cluster', often reserving – sometimes actively defending – exclusive use of the term for their version alone, no matter what the application or purpose.

Investigations of industrial clustering behaviour among firms start from either micro or macro perspectives (Belussi 2006; Steiner 2006). Both approaches account in different ways for many of the same strong, frequent, and often striking interdependencies firms enjoy with one another.

Interdependent firms typically enjoy fairly close proximity, the result of shared access to one or more uniquely supplied local resources (Gallaud and Torre 2004; Feldman *et al.* 2005a; Torre and Rallet 2005). These include the availability of valuable infrastructure services that supply indirect (and underpriced) inputs, various economic or technological spillovers from other firms and regional institutions, and embedded social advantages that work almost invisibly to reduce transaction costs and improve the efficiency of trust-based market and governance systems (Garnsey and Longhi 2004). These may arise in very subtle ways, perhaps initially from historical accidents of 'path-dependent' advantage, where the sustained velocity of initial regional advantage continues to propel the long-term success of firms in one or several linked industries (Martin and Sunley 2006).

Nevertheless, although clusters provide a provocative and holistic perspective on interdependence, the research and policy activity on clusters to date has probably raised more questions than it has answered (Lagendijk and Charles 1999; Benneworth and Charles 2001; Malerba 2002). Some of the most important of these are:

1 the sources of technological externalities that drive increasing returns in industry clusters
2 the role of social and cultural versus economic factors in determining such externalities
3 the role of proximity as an influences on externalities
4 the prospects for leveraging technological externalities through policy interventions
5 the implications of spatially targeted development policy for the growth prospects of lagging regions.

In reality, of course, applying any cluster classification is not an easy task (Coe *et al.* 2007). For example, Silicon Valley in California can be seen as a high-technology innovative cluster, a flexible production hub-and-spoke cluster and a state-anchored district. Silicon Valley is not just described in terms of local interactions but also in its global connections with partners, suppliers, and customers. Furthermore, some places are constituted by clusters that are hybrid forms merging the characteristics of clusters. Therefore, clusters are not necessarily only shaped by local forces but are also conceptualised as being defined by patterns of both local and non-local interaction (Malmberg 2003; Bathelt *et al.* 2004).

Clusters and policy

Why should governments have a role to play in strengthening or facilitating the emergence of strategic and developing clusters? In practice, four rationales for government action are apparent. The first is classical market failure rationales, namely creating favourable framework conditions for the smooth functioning of markets and the externalities associated with investment in R&D (Audretsch and Feldman 1996a; Henderson *et al.* 2001). The second is knowledge creation in global-local networks (Jacobs and de Man 1996; Dyer and Nobeoka 2000; Edwards 2002; Bathelt *et al.* 2004; Audretsch and Lehmann 2006). The third is derived from the fact that government itself is an important player in some parts of the economy (Malecki and Nijkamp 1988; Etzkowitz 2006; Etzkowitz *et al.* 2007). The final rationale is directly related to the innovation systems. These four rationales are not specific to cluster policymaking and can be adopted in other fields (Lagendijk and Charles 1997; Roelandt *et al.* 1999; Benneworth and Charles 2001; Edquist 2001; Benneworth and Henry 2003).

Clusters and policymaking: impact on industry

There has been a movement towards consensus as the literature deepens, with most progress occurring in the area of methodologies for identifying and documenting clusters.

Policies differ based on varying definitions of clusters, possible levels of analysis, and the degree to which clustering constitutes the central focus. Industry cluster principles are often used to improve the implementation of traditional development to leverage synergies arising from economic and spatial interdependence between economic actors.

Government inevitably plays a variety of roles in an economy (Lagendijk and Charles 1999; Tichy 1998; Drejer *et al.* 1999). Its most basic role is to achieve macroeconomic and political stability (Kuncoro 2002; MOSR 2002; Gammeltoft and Aminullah 2004). A second role of government is to improve general microeconomic capacity through improving the quality and efficiency of general purpose inputs to business and institutions identified in the diamond theory, such as an educated workforce, an appropriate physical infrastructure, and accurate and timely economic information. This is the case for Indonesia (Kuncoro 1996; Hill 1997). A third role of the government is to establish overall microeconomic rules and incentives governing competition that will encourage productivity growth (Porter 1998c; Burgess and Venables 2004). A fourth role of government is to develop and implement a positive, distinctive, long-term economic change process or action programme which mobilises government, business, institutions, and citizens (Etzkowitz 2006; Etzkowitz *et al.* 2007). Economic progress may be often thwarted by inaction and by a lack of consensus on what steps are necessary. Thus, a healthy change process must involve all the key constituencies and must rise above the interests of any particular administration or government (Lagendijk and Charles 1999; MOSR 2002). Ideally, such an action programme will occur not only at the national level but also at the level of states and cities (Drejer *et al.* 1999; Porter 2000; Etzkowitz *et al.* 2007).

It is claimed that clusters offer opportunities to improve productivity and support rising wages, even those that do not compete with other locations (Dicken 2000; Von Hippel 1998; Feldman *et al.* 2005a). Thus a cluster not only contributes directly to national productivity but also can affect the productivity of other clusters (Feldman 2000; Breschi and Lissoni 2001; Audretsch and Feldman 2003; Iammarino and McCann 2006). So traditional clusters, such as agriculture, wine, etc., and clusters in less-developed regions should not be abandoned; rather, they should be upgraded (Bell and Pavitt 1993; Romjin 2002). Efforts to upgrade clusters might have to be sequenced for practical reasons, but the goal should be to eventually encompass all of them. Upgrading in some clusters will reduce employment as firms move to more productive activities, but market forces – rather than government decisions – should determine which clusters will succeed or fail (Tichy 1998; Benneworth and Henry 2003; Tallman *et al.* 2004; Etzkowitz and DeMello 2004).

Accordingly, governments should reinforce and build on established and emerging clusters rather than attempt to create new ones (Tichy 1998; Held 1996; Feldman and Francis 2004). New industries and new clusters emerge from established ones as economies develop (Moulaert and Swyngedouw 1991; Sugiyama and Fujimoto 2000; Saxenian 2002; Ernst and Kim 2002). Advanced technological activities are more likely to develop where there is already a base of less sophisticated activities in the field. Clusters form when there is a foundation of locational advantages on which to build. Most clusters form independently from government and sometimes in spite of it. There should be some seeds of a cluster that have passed a market test before cluster development efforts are justified (Benneworth and Henry 2003).

The process of cluster upgrading involves recognition that a cluster is present and then removing obstacles, relaxing constraints, and eliminating inefficiencies that impede productivity and innovation in the cluster (Tichy 1987, 1998). Constraints include human resource, infrastructure, and regulatory constraint. Some of these can be addressed to varying degrees by private initiatives. Other constraints, however, are the result of government policies and institutions and must be addressed by government (Moulaert and Swyngedouw 1991; Edquist 2001; Etzkowitz 2006). Porter (2000) argues that, ideally, all government policies that inflict costs on firms without any compensating, long-term competitive or social value should be minimised or eliminated. As a consequence, upgrading clusters requires going beyond improvements in the general business environment to see how policies and institutions affect particular concentrations of related firms and industries (Kuznetsov and Sabel 2007).

Thus, governments should focus their cluster policies on externalities, linkages, spillovers, and supporting institutions to modern competition (Martin and Sunley 2003; Kuznetsov and Sabel 2007; Asheim *et al.* 2007). By grouping together firms, suppliers, related industries, service providers, and institutions, government initiatives and investment can address problems common to many firms and industries without threatening competition (Morgan 1997b; Asheim *et al.* 2007).

A government role in cluster upgrading, then, will encourage the building of public or quasi-public goods that significantly affect many linked businesses. Additionally, government investment focused on improving the business environment in clusters, other things being equal, might well earn a higher return than those aimed at individual firms or industries or at the broader economy (Pavitt 1987a, 1987b; Jacobs and de Man 1996; Harada 2003).

The appropriate government priorities change as a cluster matures and develops, and as its sources of competitive advantage shift. Early priorities involve improving infrastructure and eliminating diamond disadvantages. Later priorities revolve more around constraints and impediments to innovation.

A government role in cluster development and upgrading should not be confused with the notion of industrial policy. In fact, the intellectual foundations of cluster theory and industrial policy are fundamentally different, as are their implications for government policy (Malecki and Nijkamp 1988; Asheim and Cooke 1999). Industrial policy rests on a view of international competition in which some industries offer greater wealth-creating prospects than others (Dicken and Malmberg 2001; Dicken 2003; Coe et al. 2007). Desirable industries should be targeted for support. Industrial policy sees competitive advantages as heavily determined by increasing returns to scale (Maher 2001). Hence, governments should nurture high-priority emerging industries until they reach a critical mass through subsidies, elimination of destructive or wasteful internal competition, selective import protection and restriction for foreign investment (Humphrey et al. 2000; Edwards 2002; Nicholas and Maitland 2002). Subsidies and suspension of internal competition should concentrate on scale-sensitive areas such as R&D and facilities investment (Porter 2000; Bryson and Henry 2005).

Regularly, industrial policy tends to centralise intervention decisions at the national level. Through such intervention, government attempts to tilt competitive outcomes and international market share in a nation's favour (Roelandt et al. 1999). In contrast, cluster theory could hardly be more different (Benneworth and Charles 2001). The concept of clusters rests on a broader dynamic view of competition among firms and locations, based on the growth of productivity-interconnections and spillovers within a cluster, which are often more important to productivity growth than the scale of individual firms (Markusen 1996; Asheim and Isaksen 2002).

All clusters can improve their productivity including traditional clusters. Rather than recommending the exclusion of foreign firms, cluster theory calls for welcoming them (Porter 2000; Gordon and McCann 2000; Humphrey and Schmitz 2000). Foreign firms enhance cluster externalities and productivity, and their activities in a nation or state contribute directly to local employment and investment. Rather than advocate blocking imports, cluster theory stresses the need for timely and steady opening of the local market to imports that boost local efficiency, provide needed inputs, upgrade local demand conditions, and stimulate rivalry (Schmitz 2000; Steiner and Hartmann 2006; Feldman et al. 2005b).

Clusters provide a way of organising thinking about many policy areas that go beyond the common needs of the entire economy. Furthermore, cluster-based thinking can help focus priorities and guide policies in science and technology,

30 Theoretical framework

education and training, export and foreign investment promotion, and a wide variety of other areas (Steiner 1998). A location's best chance of attracting foreign investment and promoting exports, for example, is in existing or emerging clusters.

A cluster orientation highlights the fact that more parts of government have an influence on competitiveness than are normally recognised, particularly within government itself. Cluster theory makes the impacts of policies on the competitive position much clearer and more operational. Effective solutions often require different parts of government to collaborate.

Additionally, to complement policy attention at the economy-wide level, cluster policy highlights the important role of the government at several geographic levels (Malerba 2005; Breschi and Malerba 2005; Feldman 2005; Coe et al. 2007). The traditional focus of economic policy has been at the national level, and many aspects of the general business environment are best addressed there. Recently, globalisation has focused attention on worldwide multilateral institutions (Dicken and Lyold 1990; Dicken and Malmberg 2001; Narula and Dunning 2002; Dicken 2003). However, state, metropolitan, region, and local governments also have an important influence on the general business environment in a location. At the cluster level, these influences often are dominant, and clusters should represent an important component of state and local economic policy (Roelandt et al. 1999).

In practice, the cluster policy approaches of countries differ (Malmberg and Maskell 1999a, 1999b; Van Den Berg et al. 2001a, 2001b). One fundamental difference relates to the distinction between a bottom-up and a more or less top-down approach (Boekholt and Thruriaux 1998).

The bottom-up approach focuses on fostering dynamic market functioning and removing market imperfections. The starting point lies in market-induced initiatives, with the government acting as a facilitator and moderator but not setting national priorities. Countries that have adopted such an approach include the Netherlands and the USA. Within the Netherlands, the Technology in Society programme aims to employ technology to address societal problems. The programme brings together a wide range of interested parties in a consultation process where issues such as crime prevention and elderly care are discussed (Boekholt and Thruriaux 1998; Van Den Berg et al. 2001a, 2001b).

In the top-down approach, government (in consultation with industry and research agencies) sets national priorities, formulates a challenging vision for the future and, prior to initiating the dialogue process, decides on the actors to be involved (Boekholt and Thruriaux 1998; Van Den Berg et al. 2001a, 2001b). Once national priorities have been set and the dialogue groups implemented, the clustering process becomes a market-led process, with little government intervention. For instance, the Dutch government has established four centres of excellence that aim to stimulate cooperation between companies, academics and research organisations in the area of ICT (information and communication technology).

What country strategies can be discerned in cluster-based policy? Policy researchers such as Boekholt and Thruriaux (1998), Held (1996) and Porter (1998b) point to the various roles for government in cluster-based policy:

Theoretical framework 31

1 Establishing a stable and predictable economic and political climate
2 Creating favourable framework conditions for the smooth and dynamic functioning of markets (infrastructure, competition policy and regulatory reform, provision of strategic information)
3 Creating a context that encourages innovation and upgrading by setting a challenging economic vision for the nation or region
4 Raising awareness of the benefits of knowledge exchange and networking
5 Providing support and appropriate incentive schemes for collaboration and initiating network brokers and intermediaries to bring actors together
6 Acting as facilitator and moderator of networking and knowledge exchange
7 Acting as a demanding customer when addressing needs
8 Facilitating the informal and formal exchange of knowledge
9 Setting up competitive programmes and projects for collaborative R&D
10 Providing strategic information (technology foresight studies, strategic cluster studies)
11 Ensuring that public institutions (especially schools, universities, research institutes) cultivate industry ties
12 Ensuring that rules and regulations maximise flexible adaptation to changed market conditions, to stimulate innovation and upgrade processes

Nevertheless, there are some bottlenecks in cluster studies and cluster policymaking challenges:

1 In addition to statistical analysis, most countries combine their statistical cluster analysis with qualitative and monographic cluster case studies. One of the major disadvantages of case studies is that the approach is intrinsically qualitative. A quantitative approach is needed to map production relations, innovative networks and clusters of economic activity. Combining the qualitative cluster studies with input–output analysis can considerably reinforce the results. The dynamics of innovation identified through statistical analysis can only be meaningfully interpreted in combination with qualitative insights that arise from monographic case studies.
2 From the perspective of policymaking it is possible to identify several bottlenecks or pitfalls. Held (1996), Dunning (1997), Porter (1998b), and Roelandt *et al.* (1999) show that an awareness of these can prove advantageous when designing comprehensive cluster-based policy.
3 Cluster policymaking should not result in industrial policymaking in disguise. The creation of clusters therefore cannot be government-driven but instead should result from market-induced and market-led initiatives. For the same reason, government policy should not be strongly oriented to directly subsidising industries and firms, or to limiting rivalry in the marketplace.
4 Cluster policymaking also implies a shift away from direct intervention towards indirect inducement. Public interference in the marketplace can only be justified in the presence of clear market or systemic failure. Even if

32 *Theoretical framework*

clear market and systemic imperfections exist, it cannot necessarily be concluded that government intervention will improve the situation. Government should not try to directly lead or own cluster initiatives. Rather, the role of government should be to work as a catalyst and broker, bringing actors together and supplying the necessary support structures and incentives to facilitate clustering and innovative processes.

5   Which clusters should government focus on? Cluster policy should not ignore small and emerging clusters; nor should it focus only on existing or classic clusters. Moreover, clusters should not be created from scratch in declining markets and industries. The cluster notion has sometimes been appropriated by (industrial) policymakers and used as a rationale to continue with defensive industrial policymaking under a less traditional guise.

A final complication concerning clusters, both in methodological and policy terms, arises from changing specialisation patterns on a worldwide basis (a difficulty which also applies to sectoral analyses) (Lagendijk and Charles 1999). A trend towards growing specialisation among OECD countries, both within the same industry groups and clusters in different countries, has been observed (OECD 1999a, 1999b; Benneworth and Charles 2001; Edquist 2001). This implies that the aforementioned growing importance of networking between dissimilar and complementary firms with different specialisation patterns also has an important international dimension. As a consequence, the innovation systems and specialisation patterns of the same clusters (operating in value chains producing products and services for the same end product markets) within different countries can differ significantly in their institutional settings and innovating performance. This makes identifying best practices or optimal incentive structures in innovation systems almost tautological. International comparative research in this field can reveal the critical factors of these diverging strategies.

The risk of regional clusters

Recently the promotion of industrial clusters has been regarded by regional policymakers and development economists alike as one possible alternative to the traditional regional economic policy which concentrates on supporting individual sectors or firms (industrial or strategic targeting) (Humphrey and Schmitz 2000; Iammarino and McCann 2006). Cooperation within geographically concentrated networks of firms and research institutions, as well as governmental institutions, is thought to make lagging regions more competitive, and to accomplish sustainable economy amidst the dangers of globalisation (Amin and Thrift 1995; Wolf 2004). Prominent examples of well-functioning clusters in fast-growing regions are Silicon Valley, Little Italy, and the automotive cluster in the Austrian province of Styria (Steiner 1987, 2006).

These policies, however, ignore the fact that there are some dangers and instabilities for the region introduced by the formation of certain types of clusters (Steiner 2006). Two types of risk can be distinguished: structural risk concerns the long-term development of the region, whereas cyclical risk is associated with regional growth fluctuations in the short and medium terms (Tichy 1987; Ernst and Kim 2002; Iammarino and McCann 2006).

Regional policymakers follow cluster strategies in order to make their region more competitive and thereby accomplish sustained economic growth. But when pursuing such policies, they should also be aware of the potential drawbacks that are associated with regional clusters (Cooke 2001, 2004; Feldman *et al.* 2005a; Asheim and Coenen 2006).

Nevertheless, for this research, it is not the goal to carry out a very stringent and systematic analysis but rather to point out the problems, discuss some related aspects, and in particular suggest a methodology that appears useful for conducting further research on this matter.

The first type of risk, structural risk, has been discussed to some extent by the Austrian regional economist Tichy (1987, 1998). It is related to the long-term consequences for a region dominated by an industrial cluster. The second type of risk, often named cyclical risk, refers to economic stability which is, besides sustained growth, a common goal of regional policy.

History shows that the regional concentration of resources in one industry or product group, as may be implied by the formation of a cluster, bears the risk that a permanent decline in that industry or product group may bring down the whole region (Storper 1995; Scott and Storper 1987). Detroit is probably the best known example of a metropolitan region that became a so-called 'old industrial area' because of the crisis in the US automobile cluster, with activities heavily concentrated in and around Detroit (Klepper 2001, 2002, 2004).

A similar crisis occurred in the Austrian region of upper Styria. Its steel cluster, composed mainly of nationalised firms which were quite successful in the 1950s and 1960s because of process innovations, began declining in the 1970s before it finally broke down completely in the 1980s (Steiner 1998, 2006; Tichy 1998). These clusters, once dynamic, successful, and innovative, could not remain at the competitive edge of their respective industries. They both failed to adjust to changing production regimes (Steiner 2006; Audretsch and Lehmann 2006).

Tichy (1987, 1998) examines some factors responsible for what he calls ageing, or petrifaction, of clusters. At the core of his analysis is the theory of regional product cycles. This theory states that goods are created and initially produced in agglomeration because of informational advantages in the face of high uncertainty, concentrated specialised demand, and the availability of skilled labour. Later on, when production processes become more standardised, production moves to the periphery, where labour is less skilled but cheaper and where economies of scale are utilised (Vernon 1996).

It is at this stage, Tichy (1987, 1998) claims, that the region specialises, and networks or clusters are formed. But once the products reach the final phase of their life cycle and their competitive advantage is shrinking, concentration of the number of firms and activities sets in. Thus the networks become smaller and the information flow is reduced. Smaller clusters are less likely to stimulate innovations, so the region finally degenerates into an old industrialised area without any endogenous potential to regain its competitiveness.

A key issue in this respect, which has also been mentioned by Markusen (1996, 1999), is the link between the success of a cluster, its increasing specialisation, and the subsequent tendency to become a 'closed system'. As a result, the information necessary for rapid adaptation to market changes may not reach the cluster firms. Furthermore, success and specialisation in one industry may impede the development of other sectors in the region or even drive out firms, preventing the diversification of the regional economy (Cooke 2006; Nooteboom 2006; Malmberg and Power 2006).

This may occur when public funds (for example, for education, research, or infrastructure) are allocated according to the importance of regional industries, so that clusters receive a larger share. Industries outside the cluster will be put at a disadvantage, particularly if these publicly produced goods are indivisible. A cluster-oriented policymaker may also allocate more direct and indirect subsidies to cluster industries. Additionally, successful industries will attract workers by offering higher wages which may cause deterioration in the human resources of other, less productive, industries. As a result, the risk of becoming a problem increases once the cluster declines, together with greater exposure to cyclical instability.

Regional policymakers are not only concerned with regional growth prospects, but also try to avoid unnecessary fluctuation in business activities, because they cause inefficiency and welfare losses (Benneworth and Henry 2003). Among the resources of instability in regional economic activities the national business cycle is the most important, but because of increasing globalisation the region can also be affected by unexpected events outside the nation, for example, changes in international energy prices. Because business cycles are unevenly distributed among industries, the region's industrial structure (that is, its mix of different industries), determines the level of instability for the region as a whole (Benneworth and Charles 2001). Consequently, the search for the optimal industrial mix has been ongoing in the literature for some time.

Since the earliest economic history, clusters of activity have proved indispensable for economic success (Venables 2003). Chinese silk or Persian carpets; English machines or German dies; Swiss watches; all are examples of regional specialisation in the form of clusters – regional networks of interlocking activities, based on sophisticated systems of division of labour among skilled specialists. Today, science-based industry, concentration on core activities, and outsourcing are increasing areas with deserted plants and unemployed workers. This illustrates the end of at least some clusters that began in such a promising way. To some extent the same factors that cause the success of young clusters may also cause the demise of old clusters. Therefore, it is necessary to find those factors in order to find a remedy to prevent clusters becoming old, or to effect their rejuvenation.

That some types of clusters have a limited life span needs no long explanation (Benneworth and Charles 2001). Raw material based clusters necessarily face problems when deposits are depleted or the materials are no longer in demand. The coal and steel regions of Europe are some of the best examples of petrified clusters and the transition of formerly dynamic and rich areas into problem areas. Product-based clusters may face similar problems, as the goods produced in the cluster age and are no longer in demand. Clusters based on local skills age when machines render skills superfluous or methods are developed to produce the goods with unskilled workers.

The examples show that clusters can grow old and petrify. But must they go to ruin? Some examples demonstrate that even raw material clusters need not age; they can transform into skill clusters or process clusters. London as a financial cluster is a good example of a successful transformation (Steiner 1998). The city no longer depends on the raw material that gave rise to its capital status; London now supplies skill-based services to foreign capital, to borrowers as well as to lenders. Examples of production clusters that have so far been able to stave off ruin are machine tools in Baden-Württemberg, motor cars around the Great Lakes, or watches in Switzerland. All of them have been 'reclustered' in one way or another.

The evidence suggests that clusters tend to age, to become inflexible, and, in the worst cases, unable to adapt to a new environment. Why this does happen? What prevents clusters from adapting to new conditions? Why can they not adapt their skills when the old specialisation can no longer meet demand? The theory of the regional product cycle (Tichy 1987, 1998) can provide an answer. The motto could be 'learning new skills is easy, unlearning old habits is tough' (Lazonick 1992; Nonaka *et al.* 2000).

The theory of the product life cycle (Vernon 1996) says that goods are created in agglomerations, because of their informational advantages and the concentrated demand. In its early life such a good is produced in the agglomeration in which it was developed. As long as the product is still in its development phase, the production processes are not yet standardised and can afford the specialised skills of the region.

A most important lesson is that a cluster is more likely to become a problem area the more successful it is. The more successful a cluster, the more it specialises and concentrates on the relevant product or process, the more it attracts matching skills and drives out others; as networks become more specialised, they lose their ability to transfer other information which would be necessary to develop new products.

Clusters are therefore in danger of petrifaction, especially if they are successful (Belussi 2006). The more the cluster's firms concentrate and the more the networks shrink, the more real this danger becomes. A cluster is certainly more dynamic and less likely to petrify than a vertically integrated firm, but old clusters tend towards vertical integration, in an unattainable desire to survive by utilising economies of scale and to save on the cost of inputs (Asheim and Gertler 2005; Asheim *et al.* 2007).

Tendencies for petrifaction of clusters may arise internally as a result of the dynamics of the regional products cycle or externally as a result of shrinking markets for their products (Asheim and Coenen 2006). The latter may be caused by the fading away of former locational advantages, by new technologies, or by shifting demand. In all cases, a sclerosis of institutions and persons sets in, of entrepreneurs as well as of workers, of bureaucrats as well as labour unions (Audretsch and Lehmann 2006). They all tend to do whatever they can to sustain the old structures in the short run, thereby destroying the cluster's basis in the long run. They do not want new people to join their sclerotic club, and thereby they prevent new information and new network nodes (Bathelt 2006).

Clusters are an important element of economic competitiveness and their importance will increase even more in coming years. New management strategies and government attempts to build national innovation systems both imply concentration on a more limited number of activities (Malerba 2005; Metcalfe and Ramlogan 2006). In the short and medium term this will increase competitiveness, although at the cost of potential greater risk, resulting from the reduced portfolio.

In the long run, however, serious problems may arise if countries do not succeed in preserving the flexibility of the clusters and a high density of diversified information within the cluster regions. Policymakers for industry and technology must be aware of this problem. However, this does not demand more intervention, but better intervention, with a sense of the long-term consequences.

A policy that is restricted to rejuvenating old sterile clusters normally comes too late. Policy should try to prevent clusters from becoming sterile, it should help new clusters to form, and it should watch that newly formed clusters are constructed in a way that promise growth potential, flexibility, and a long life (Cooke 2004; Asheim and Coenen 2006). Many clusters arise out of indigenous market forces, but others are the products of policy. Science policy, research policy, technology policy, and economic policy increase focus on the promotion of clusters, and if they act unanimously, they have a good chance of succeeding (Edquist 2001; Malerba 2002; Asheim and Gertler 2005). In some cases it is a distinct policy goal to create a cluster with the help of special institutions (for example, science parks) and project managers, or, at least, to promote its formation (Cooke 2001; Etzkowitz 2006).

An already proven strength of firms and research activities in the planned cluster, and a reasonable number of nodes, are both preconditions to secure the necessary mix of information density and output diversity. Another important point is novelty, that is, innovative power (Cooke 2001, 2004; Ernst 2000; Feldman 2000; Parrilli 2006). The product or the process at the centre of the cluster should be at an early stage of development, at a stage in which several different routes to success are still open, and following a sector-based industry approach: the strengths should cut through different industries, based on knowledge cutting through different branches and interaction among them (Jacobs and De Man 1996; Asheim and Isaksen 2002; Gordon and McCann 2005). Therefore, those clusters that rely on cross-section technologies or new process technology rather than on new products have a better chance (Asheim *et al.* 2007).

Clusters in developing countries and FDI

Based on previous research on clusters in the developing countries, the outcomes of cluster development discussed in the literature mainly concern the traditional sector at SME (small- and medium-sized enterprise) level (UNCTAD 1998; Altenburg and Meyer-Stamer 1999; Bell and Albu 1999; Meyer-Stamer et al. 2001; Giuliani 2006). In relation to this research, the literature on clusters in developing countries has yet to highlight the hi-tech industry such as the automotive industry. As a result, most of the literature on the automotive cluster has been derived from the mature cluster in developed countries whereby the automotive industry has played the role of a parent company, not of a subsidiary (Klepper 2001, 2004; Cantner et al. 2004; Tallman et al. 2004; Boschma and Wenting 2006).

In order to attract foreign direct investment (FDI), many developing countries have enacted significant policy reforms. They went through what are known as the first and second generations of investment promotion policies. In the first generation of investment promotion policies, many countries adopted market-friendly policies. They liberalised their FDI regimes by reducing barriers to inward FDI, strengthening standards of treatment for foreign investors, and giving a greater role to market forces in resource allocation.

Furthermore, Hanson (2001) reported that developing countries at all levels of development have created a policy infrastructure to attract multinational firms. For example, many developing countries have taken measures such as expediting the approval process, removing restrictions on the repatriation of profits, strengthening their standard of property rights, providing liberal tax incentives, and allowing foreign participation in the privatisation of state-owned enterprises. Given all these changes and reforms, one should expect a significant increase in the rate of FDI flows into developing countries, yet it is not happening.

On the contrary, the continued capital flight from developing countries demonstrates that the private investment response to these reforms has so far been disappointing. This clearly calls for new research that can generate new investment promotion policies. However, the concept of clusters of complementary firms as a determinant of attracting FDI, while present in policy debate, has not received much attention in the economic literature.

In developing countries, clusters were thought to be a viable way to foster the development of small local (informal) industry and to eliminate the growth constraints of small realities (Schmitz 1982, 1995, 2000). Geographical and sectoral concentration (i.e. clusters) does not necessarily provide any beneficial effect unless an active process of inter-firm cooperation and productive competition has been set up, especially in critical situations.

The existing literature, in trying to explain the low capital flow in developing countries, has not focused on the importance of dynamic clusters or analysed the ability of governments in these countries to attract clusters of foreign investment. For example, Rodrik (1991) has focused on policy uncertainty and has pointed it out as a possible cause of under-investment in developing countries. He developed a model with no emphasis on linkages which ties policy uncertainty to the private investment response. Focusing on linkages, where each investor makes its investment decision based not only on its own perception of policy reform, but also on the investment decision by other investors, makes a crucial difference with Rodrik's model. It leads to a different set of results.

Rodriguez-Clare (1996) explored how multinationals through FDI affect underdeveloped regions via the generation of linkages. However, he did not examine the mechanism by which underdeveloped regions can attract them. Additionally, the contribution of FDI to knowledge spillover in a region has been gathered from previous studies (Porter 1990; Dunning 2000a, 2000b; Benneworth and Charles 2001; Narula and Zanfei 2005). However, none of these studies examined the role of automotive MNEs in developing the cluster, particularly in the context of developing countries. Therefore, the literature lacks significant presence regarding FDI in the automotive sector.

In developing economies, the first systematic effort to analyse industrial clusters has followed a common approach based on the concept of collective efficiency. This concept was firstly introduced by Schmitz (1982, 1995, 2000) in an attempt to define a methodological approach to assess the functioning of clustered firms in such countries. This focus was stimulated by the success stories of Italian industrial districts in the 1970s and 1980s that generated the interest in this new model of industrial organisation (Piore and Sabel 1994).

In developing countries, therefore, clustering was thought of as a viable way to foster the development of small local (informal) industry and to eliminate the growth constraints of small realities (Schmitz 1982: 9): '...such clustering opens up efficiency gains which individual firms can rarely attain. The concept of collective efficiency is defined as the competitive advantage derived from local external economies and joint action' (Schmitz 1995: 530). In this approach, geographical and sectoral concentration (clustering per se) does not necessarily provide any beneficial effect unless an active process of inter-firm cooperation and constructive competition has been set up, especially in critical situations.

More recent contributions have started emphasising the role of external sources of knowledge in rejuvenating local knowledge flows and on this point there is growing concern about finding a theoretical approach that could link local agglomerations such as clusters of small and medium enterprises and the global economy: as already noticed by Humphrey and Schmitz (2000: 14) the cluster literature emphasises the need to improve cooperation and local governance. Even the resources for product and functional upgrading are seen mainly to come from within localities. Links with the wider world are frequently acknowledged, but they are weakly theorised.

As a consequence, the focus of analysis has shifted from local cooperation dynamics to local-global links and has increasingly adopted the 'Global Value Chain' approach (see Gereffi 1994). Different studies, both theoretical and empirical, (see, among others, Gereffi 2001; Humphrey and Schmitz 2000; Power and Lundmark 2004; Rasiah *et al.* 2008) now place emphasis on the role of global foreign buyers as main carriers of knowledge flows towards the cluster. Buyers exert on local firms a quasi-hierarchical form of governance which is aimed at producing product and process upgrading. In fact, there is substantial evidence of increased production capabilities (in the sense given by Bell and Pavitt 1993) in local firms that upgraded the quality and the pace of their production (thus reducing time to market and time to order) (Humphrey and Schmitz 2000).

Consequently, since clusters have life cycles, regions must be continually vigilant, scanning market and technological trends (Pavitt 1987a, 1987b; Patel and Pavitt 1998; Cooke 2001; Asheim and Isaksen 2002). It is important for successful clusters to have contingency plans for changes in consumer tastes and demand, and for new technologies that might result in market shifts (Pries and Schweer 2004; Garnsey and Longhi 2004). The cluster, if well organised and self aware, can become the mechanism that gathers information, predicts shifts, and finds new opportunities (Breziz and Krugman 1993; Fujita et al. 1999; Fujita and Krugman 2004).

Knowing and understanding clusters is of value to a region only if that knowledge leads to actions that grow economies and raise standards of living. Unfortunately, there is no single recipe for developing countries to follow that will meet the needs of all clusters, which embody many types of systematic relationship and kinds of industries (Fujita et al. 1999; Fujita and Krugman 2004). But there is a menu of actions from which to choose. The choices developing countries make depend on many factors, including geography, stage of development, resource constraints, special societal needs, cluster priorities, market imperfections, and local preferences.

Conclusion: bringing the strands together

There are benefits that spill over from clusters along with the risks which might arise. While the success of an individual firm may depend on its ability to protect its own technological advances, new products, or designs, firms press competitors within the clusters to continually improve and innovate in order to maintain their advantages over imitators.

The initial development of district or cluster capabilities can be enforced by the development of an institutional system conducive to growth, where network benefits will emerge, and extensive cooperation is promoted among the local agents

Nevertheless, there is a danger of seeking to replicate elsewhere what has been done successfully in one location because of the existence of territorial specificity. This is not a linear process because the path dependent nature of local institutions and a weak innovative system can bring about a stationary situation.

Accordingly, each of the important elements in the cluster studies framework and relevant context for this thesis will be explained as follows:

1 Connections

The most successful clusters build mechanisms that can speed the movement of ideas, innovations, and information from firm to firm throughout the economy.

The mechanism and entities for collecting and disseminating knowledge – the gatekeepers, brokers, and intermediaries that encourage and facilitate all forms of associate behaviour – provide the value embodied in cluster development that is so important to cluster competitiveness.

2 Networking and networks

The single most important operating principle of competitive clusters is the ability to network extensively and form networks selectively. Networking is the process that moves and spreads ideas, information, and best practices throughout a cluster and imports them from other places.

3 Competencies

Although many factors affect the competitive advantages of clusters, none is as important as the competencies they embody. Learning and knowledge transfer represent the lifeblood, and skilled labour the gene pool, of clusters.

4 Industry leaders

Behind every successful cluster is a group of innovative firms led by people who value learning, are committed to their community and are willing to work towards a collective vision for their industry. These leading companies may have a niche or rapidly growing market that is not threatened by competition, or they may face such intense global competition that benefits of mutual support and learning outweigh concerns about confidentiality. The key to building and sustaining a cluster organisation often rests with the support of these benchmark companies.

5 Knowledge

Successful regions are home to institutions, individuals, and organisations that serve as storehouses and disseminators of both documented and undocumented knowledge. The knowledge resides in research and technology centres and their staff, educational institutions and their faculty, and companies and their employees.

6 Barriers facing clusters in developing countries

Various historic under-investments limit clusters in developing countries from gaining new, or holding onto existing, competitive advantage. Most can be traced to a weak infrastructure; lack of access to capital, technology, innovation, and capital; regional insularity and isolation; low educational levels and a low-skilled work force; absence of talent; and an overly mature or hierarchical industry structure. Social exclusion exists in places with large and isolated underprivileged and undereducated populations; technological exclusion exists in places with poor access to sources of technology and benchmark companies; and economic exclusion is a result of weak links to benchmark regions and markets.

7 Deficits in physical infrastructure

Infrastructure deficits create an uneven playing field for regions that inhibit capital investment. Locations where the transportation of goods and people are costly and less frequent pose even more severe handicaps on regions that are difficult to remedy. As logistics become more important to customers, poor access to transportation becomes more of a barrier.

8 Weak technology institutional structures

Clusters depend on regional institutions for a variety of things they cannot do internally or get from other companies. They use regional institutions for information about, and help with, advances in technologies, economic scans, brokering, and education and training at all levels. Nearly all regions have an institutional framework for education and training and for some services, but few have an explicit economic development focus. Nor do they have the resources and expertise to target clusters; thus, few have become the centres of excellence that attract talent, resources, and other companies.

In summary, the literature so far reviewed is heading, both in the context of advanced countries and of developing countries, towards a better understanding of the mechanisms governing knowledge acquisition, diffusion and generation in clustered firms. Nevertheless, at least in developing countries, a further effort is needed to join external knowledge flows, local knowledge systems and firms' knowledge and technical capabilities into a unique framework. Missing in the literature reviewed is the distinction between the knowledge that is accessible to firms (coming from fairs, specialised journals, customers, business associations) and that which is (potentially) absorbed by the firm. Clearly, the links among firms (both internal and external to the cluster) are important sources of knowledge when firms are capable of capturing that knowledge in a way that leads to adaptive change or innovation.

Even though no 'recipes' can be formulated to guarantee a successful outcome, the review of both theoretical and empirical contributions regarding advanced and developing countries suggests that (at least in the latter) a further step is required in the understanding of the economics of knowledge absorption and generation at a firm level of analysis. This cannot be pursued without a clear framework that takes into consideration the role firms play within local and external (national or international) networks. For the Indonesian automotive cluster, it is a fundamental question whether or not Toyota and Honda (both MNEs) and their local subsidiaries have gained from having a presence in the cluster.

42 Theoretical framework

Accordingly, clusters and the relevant concepts can only be explained and uncovered through detailed empirical work. My empirical research into the automotive cluster in Indonesia touches on some of the key notions and relations within the cluster, but my main focus is on the way in which the leading MNEs (Toyota and Honda), through knowledge transfer, develop the growth of knowledge and networks and the way in which Indonesian industrial policy has impacted on the assets and relationships within the automotive cluster. Through such empirical work we can seek to add to the theoretical understanding of clusters in developing countries by further unpacking cluster concepts and considering the usefulness and fit of existing theories that have sought to explain clusters and the processes within them.

Chapter 2 has explored the literature regarding cluster theory and the use of cluster concepts in industrial development. It has considered how, and why, clusters have become influential in industry and development studies. In addition, this chapter has outlined the implications for cluster policy and the risks of regional clusters and the highlight of cluster studies in developing countries, mainly from traditional sector perspectives.

3 Production and knowledge transfer in Japanese automotive networks

In Chapter 2, the definition and the explanation for the functioning and development of clusters were explored, and it was understood that clusters are composed of complementary and relevant theoretical strands. Another major factor of this analysis has to be the systemic character of knowledge transfer within the automotive industry.

I now want to look at the way the Japanese multinational enterprises (MNEs) perform in the automotive industry. This insight will later play an important role in examining the Indonesian automotive clusters; particularly their effects on knowledge and technology transfer in the case of the Japanese transplants in Indonesia, and whether and how they have developed knowledge and technological capability in the Indonesian automotive cluster.

In conjunction with the internationalisation of the automotive manufacturing industry, especially in the context of knowledge transfer in economically peripheral regions, the Japanese MNEs are truly expanding their 'know-how' globally (Chen 1996; Birkinshaw 1996; Harada 2003). The Japanese automotive sector through FDI has grown increasingly important since the 1960s (Yamashita 1991; Smitka 1991; Han 1994; Rugman and D'Cruz 2002; Fujimoto 2007). Moreover, this sector is becoming increasingly expertise-intensive as a result of structural changes in automotive value chains (Dicken 1988, 2003; Klepper 2001, 2004; Coe et al. 2007). Thus looking at the learning aspect from a regional perspective (Florida 1995; Asheim 1996; Morgan 1997a), as well as from organisations (Womack et al. 1990; Wickens 1995; Steiner 2006; Poon et al. 2006; Fujimoto 2007), could be a fruitful discussion in the context of the automotive cluster.

In view of that, Chapter 3 emphasises the importance of understanding Japanese automotive MNEs in the case of Japanese transplants in Southeast Asia and globally. The Japanese automotive industry has different characteristics in its global-local interlinked network, also known as *keiretsu*[1] (a network in the network).

This chapter also highlights the relationship between different actors involved in the Japanese networks and the element of knowledge transfer from foreign direct investment (FDI) channels. The conclusion looks at how a review of the existing literature contributes to this research from the perspective of the Japanese automotive industry and its networks.

Knowledge transfer by Japanese MNEs in the automotive industry

Based on the traditional approach towards technology transfer, each activity of a company is maintained within the firm or externalised according to transaction cost criteria. By contrast, in a knowledge-based economy it is no longer possible to control each part of the value chain on the same level; hence, companies have to concentrate on some special expertise, known as 'core competences' (Nooteboom 2003; Malmberg and Power 2005).

The definition of transfer can be further complicated by the diverse channels through which it can occur. There is no best way of transferring knowledge in the automotive clusters for two reasons: first, technology does not exist in a social vacuum, but rather is 'tacit' and embodied in products, processes, and people; second, technology circulates through very diverse institutional channels or mechanisms. There are both formal and informal channels, some of which involve voluntary and international technology transfer and others do not (Odaka *et al.* 1988; Chen 1996; Liu and Dicken 2006; Fujimoto 2007).

In conjunction with the nature of the global business network of MNEs in the automotive sector, the principal channels of international technology transfer are: licensing; franchising; FDI; joint ventures; subcontracting; cooperative research arrangements and co-production agreements; export of high-technology products and capital goods; reverse engineering; exchange of science and technical personnel; science and technology conferences; trade shows and exhibitions; education and training of foreigners; commercial visits; open literature (journals, magazines, books); and government assistance programmes. (Chen 1996). Technology transfer through most of these channels is difficult to monitor. Thus, through these formal channels, technology is transferred via a market mechanism and institution along with the relevant actors involved (Schnepp *et al.* 1990; Chen 1996).

In view of that, to make research manageable, this research applies a narrower definition of formal channels, which defines technology transfer in the automotive clusters as 'a process by which expertise or knowledge related to some aspect of technology is passed from one user to another for the purpose of economic gain'.

The case studies from the automotive industry make clear that the Japanese MNEs have extended important influences upon the global automotive industry and indirectly upon governmental automotive industrial policies. It is when Japanese industrial organisation models are transferred abroad that they influence modes of technology transfer and systems for local skill formation in a particular place (i.e. automotive cluster) (Busser and Sadoi 2004). At the same time, however, Japanese industrial organisation models show a strong ability to adapt themselves to local conditions (Nonaka *et al.* 2000; Ozawa 2005; Fujimoto 2007).

The choice to study the automotive industry in terms of knowledge and technology transfer from Japanese MNEs to foreign enterprises and the Japanese influence upon skill formation systems can be argued from many perspectives (Ichiro 1991; Smitka 1991; Kenney and Florida 1993). First, the automotive industry has a wide variety of technologies in use, ranging from simple assembly and plastic moulding to state-of-the-art robotic welding technologies (Nonaka *et al.* 2000; Fujimoto 2007). Second, governments in both developed and developing countries perceive the automotive sector as an important means by which to enhance their industrial structure (Dicken 1988; Odaka *et al.* 1988; Miyakawa 1991; Mai 1991). Therefore, many countries have adopted automotive-specific industrial policies.

Additionally, the sheer size of the automotive sector makes it difficult to ignore. Since its emergence, the automotive industry has created new models in industrial organisation. The American domination of the industry – so-called 'Fordism' – continued into the 1970s (Womack *et al.* 1990). However, in that decade, fuel-efficient Japanese cars started to conquer the global market while still using mass production systems. In the 1980s, Japan witnessed a gradual shift away from mass production towards lean production systems. The rising dominance of Japanese automotive producers is illustrated by the term 'Toyotaism' that emerged in the late 1980s (Dohse *et al.* 1985; Kadoto 1985; Nomura 1993; Shimizu 1999). As a result, the Japanese automotive industry took off in the 1980s and still continues, justified by technology transfer from Japanese to foreign enterprises and Japanese influences upon skill formation systems (Nonaka and Takeuchi 1995; Nonaka *et al.* 2000; Fujimoto 2007).

There are a number of parties (i.e. actors) in international technology transfer in the automotive industry (Hieneman *et al.* 1985; Kiba and Kodama 1991; Hatch and Yamamura 1996), but more important are national governments that are concerned with how international technology transfer affects local economic development, international competitiveness, and national security (Gorgh and Greenaway 2002). In short, they are concerned about the need for, and the usefulness of, national control on either outflows or inflows of technology (Okada 1983; Chen 1996; Edwards 2002).

Along with the government, corporations and individuals who directly participate in international technology transfer in the automotive industry as either suppliers or recipients of technology are also interested in a better understanding of the field (Chen 1996; Humphrey and Schmitz 2000; Malmberg 2003; Asheim and Coenen 2006).

In view of this, to comprehend the nature of Japanese MNEs, it is necessary to distinguish the classification of the MNEs and knowledge diffusion at the beginning of their globalisation engagement. This classification is based on the Japanese MNE destination as an appropriate approach to explaining both the traditional global strategy of Japanese MNEs and the recent shift in the pattern of Japanese automotive MNEs all over the world (Dicken and Henderson 2003; Fujimoto 2007). FDI in Japanese automotive MNEs is classified in two ways: developing country-oriented FDI and developed-country-oriented FDI. For a substantial period, the Japanese economy has relied for its maintenance and growth upon external trade with foreign countries; hence, it has been an export-oriented economy. During the process, the necessity to utilise cheap labour and natural resources in neighbouring countries has been widely recognised. As a result, they have been able to establish a well-formed network of production alliances that include Indonesia.

In contrast, FDI in developed countries, such as the US and European countries, had been considered initially only as the markets for completed products rather than as the providers of resources of the sites for production plans (Dicken and Henderson 2003; Fujimoto 2007). This connotes that there was a sharp distinction between the production and marketing functions in Japanese industry; hence, production was centred in Japan and in the neighbouring Asian countries whereas marketing was focused in the developed European countries and the US.

In developed countries, on the contrary, where indigenous companies possess their own higher technology, the possession of more advanced technology appears to be critical for the success of the Japanese subsidiaries. Moreover, Japanese companies are placed at a relative disadvantage in securing low-cost labour and materials; hence, the Japanese subsidiaries in developed countries tend to equip themselves with higher technology than is the case of those in developing countries.

MNEs are major vehicles of international technology transfer in the automotive clusters in Southeast Asia as their activities cover the selection of technologies for transfer, the choice of transfer channels, and transaction pricing and contractual mechanism and negotiation (Pedersen *et al.* 2003). Consequently, they have to deal with the attempts of national governments to impose various restrictions on international technology transfer in multiple exchanges (Nooteboom 2003; Asheim and Coenen 2006; Steiner 2006). International automotive technology transfer will involve the international participation of a business and it can be examined by following sequential stages in the life cycle of the product or process development (Kiba and Kodama 1991; Szulanski 2000; Pedersen *et al.* 2003). New modes of transferring technology can result in profound institutional change, both in the parent company or the host country (Asheim and Cooke 1999; West 2000). Traditional institutions have transformed into new organisations characterised by their knowledge transfer/knowledge-sharing platforms. This kind of organisation can overcome various fundamental obstacles to the evolution of an effective environment for a knowledge-based economy (Amin 2002; Asheim and Coenen 2006).

Additionally, understanding technical change and knowledge is crucial for understanding the dynamics of knowledge-based economies and learning economies (Asheim and Cooke 1999; Asheim and Coenen 2006). Differences in sectoral industry performance and the related institutional settings particular to a country partly explain variations in economic performance (Birkinshaw 1996; Markusen 1999; Bathelt 2006), in particular, the lean production system which is the Japanese automotive industry's most important comparative advantage to transform the knowledge and technology capability of the host country. However, the lean production system cannot easily be transferred abroad. Even if the production facilities and processes are transferred, to operate effectively the organisational structure also has to be transferred (Gertler 2003; Ozawa 2005; Fujimoto 2007).

Knowledge transfer and the upgrading of productive capacity in the automotive industry is a dynamic social process that evolves most successfully in a network in which intensive interaction takes place between those producing and those purchasing and using knowledge. Furthermore, the literature on knowledge transfer highlights two essential dimensions of the learning aspect of transferring technology knowledge (Cooke *et al.* 2003; Nooteboom 2003; Asheim and Gertler 2005; Asheim and Coenen 2006):

1 The interaction between different actors in the knowledge transfer process, particularly between users and producers of the intermediate goods, and between business and the wider research community, is crucial to successful improvement (interdependency).
2 Institutions matter, because knowledge transfer processes are institutionally embedded in the setting of systems of producing (systematic character).

At the same time, the rate of specialisation is rising. Companies in the automotive industry are developing strategies to cope with their increasing dependency on their environment (Hess and Yeung 2006; Liu and Dicken 2006; Mikler 2007): for example, more flexible Japanese automotive organisational structures and the integration of various elements in the production chain through strategic alliances, joint ventures and consortia in the name of *keiretsu* (Samuels 1987; Dicken 1988; Smitka 1991; Kiba and Kodama 1991; Terry 2002). The division of labour between dissimilar and complementary firms is based on the strategic choice that firms have to make between internalising knowledge and sharing information with external actors. One way to achieve it is by forming strategic alliances. The main goal of most strategic alliances has been to gain access to new and complementary knowledge and to speed up the learning process (Han 1994; Kogut 2000; Edwards 2002).

In organisational learning in the automotive sector, technological knowledge is not the only important skill; steering skills are also important (Sugiyama and Fujimoto 2000; Steiner 2006; Fujimoto 2007). Tacit (and other forms of) knowledge are increasingly acquired within automotive firms through deliberately planned and funded activities in the form of product design, production engineering, quality control, education and staff training, research, or the development and testing of prototypes and pilot plants (Chen 1996; Larsson 1999; Dyer and Nobeoka 2000; West 2000; Fujimoto 2007).

In the automotive context, the process of R&D must be competently organised along with production systems in manufacturing (Doner 1991, Wickens 1995; West 2000; Fujimoto 2007). These processes take place within a social system; the actors of this system are interlinked actors who are based at a global-local network. Further actors of the macro-network, such as those in government, universities, and business are also playing their roles on the automotive macro strategy but not on the micro level, although they might influence the innovative environment more indirectly and in the long term (Humphrey et al. 2000; Klepper 2002; Alvstam and Schamp 2005). Therefore, global challenges between parent companies and subsidiaries in the resources devoted to such deliberate learning – or technological accumulation – have led to international technological gaps, which in turn have led to international differences in economic performance (Klepper 2004; Boschma and Wenting 2004; Cantner et al. 2004).

In view of that, tacit knowledge (underlying the ability to cope with complexity) is acquired essentially through experience, and trial and error (Nonaka and Takeuchi 1995; Nonaka et al. 2000; Ernst and Kim 2002) particularly for the expansion of the Japanese automotive industry. Particularly, the role of trust and social capital within knowledge exchange is described as a dyadic relationship between the actors who have previously exchanged knowledge (Maskell and Malmberg 1999a, 1999b). This strategy is called 'Global-Local Corporation' and works alongside the *keiretsu* (Miyakawa 1991; Kenney and Florida 1993; Rutherford 2000). Thus, managing international technology knowledge transfer can be further complicated by the diverse channels through which it can occur (Chen 1996; Fujimoto 2007).

The link to clusters is that this type of network is more likely in a territory where individuals from different firms know each other and may belong to the same industrial organisations and associations, and where new firms will already be used to the unwritten rules of conduct. Therefore, knowledge transfer is related to territory and institutions. This establishment of shared trust allows firms to have an advantage from trustful relationships without having to invest in building individual trustful relationships.

Japanese automotive production networks in Southeast Asia

The forces determining the spatial location for business and production of the automotive industry can be divided into two categories: internal and external forces. Internal forces refer to the specific factors internal to industries which cause factories in an industry to cluster or disperse, such as organisational structure and external economies of scale (Dicken and Lyold 1990). The external forces of location are those of the area that are characteristic in attracting companies in a particular industry (Han 1994).

On the other hand, factors which influence the locational shift of manufacturing companies are also divided into two categories: push factors, which make a company move from an existing site; and pull factors, which attract a company on the move (Dicken 2000, 2005; Coe et al. 2004). External forces and pull factors are likely to be related to geographical locations. Meanwhile, push factors tend to imply both geographical and non-geographical motives in industrial location.

The growth of the Japanese automobile industry in the 1980s was a breakthrough for more than the speed of its increase in production volumes. The rapid expansion of overseas production, the integration of these overseas operations in specific clusters into a global production network and the accompanying expansion of integrated global supply networks also attracted attention (Busser and Sadoi 2004; JAMAI 2008).

The rise of the Toyota-style 'lean production system' was superior to any other production system in the industry (Womack et al. 1990). The overriding sentiment in the automotive sector was that it was competitive to fight the Japanese competitors on their own terms and, since then, the Japanese model of production has become the standard (Chen 1996; Sugiyama and Fujimoto 2000; Liker and Meier 2005; Fujimoto 2007). However, it was proven that it is a mistake to treat all Japanese automakers as equal. In the late 1990s, Nissan, Mitsubishi and Isuzu were in financial difficulty, particularly in their European and North American markets (Rutherford 2000; Terry 2002).

The lean production system was the Japanese automobile industry's concern. Many studies were conducted into the Japanese approach to the organisation of production (Liker and Meier 2005; Fujimoto 2007). These studies agree that the major strengths of the lean production system are smooth production flows combined with product variety, consistently high standards and continued improvements to both the products and the production processes (Sekiguchi 1983). As the lean production system is dependent on a specific type of industrial organisation, the transfer of production technology abroad by Toyota, Honda, Nissan and Mitsubishi required the formation of local supply networks similar to those in Japan. In the late 1980s, local suppliers both in the US and in Southeast Asia were not available to Japanese final manufacturers (Rutherford 2000; Terry 2002).

In the US, the Big Three produced high ratios of parts and components in-house, while in Southeast Asian automotive clusters, technological levels of the local suppliers were of a poor standard. As a result, Japanese final assemblers requested their suppliers to invest in production facilities in the US and in several Southeast Asian countries. In this way, a number of important features of Japanese MNEs were transplanted abroad into the local cluster in the host countries (Smitka 1991).

Subsequently, Southeast Asia's early attraction for Japanese auto firms lay partly in the region's general cluster growth (Doner 1991; Rasiah 2005; JAMAI 2008) and its auto markets in particular. The ASEAN-4 automotive clusters (Indonesia, Thailand, Malaysia and the Philippines) played an important role in the industry's early stages which made Southeast Asia a logical focus of early Japanese automotive expansion. Japan's proximity to, and war-time position as occupier of, Southeast Asia, provided contacts and encouraged a view of Southeast Asia as a strategic buffer for Japanese firms (Smitka 1991; Borrus 1992; Aswicahyono and Anas 2000; Yoshimatsu 2002).

The ASEAN markets constituted a testing and training ground for Japanese vehicles and personnel preparation for penetration of markets in advanced countries (Doner 1991; Dicken et al. 1995; JAMAI 2008). In view of the auto industry's central role in Japanese manufacturing, pessimistic predictions for OECD market growth, and the assumptions of the importance of East Asia and the ASEAN-4 automotive clusters for Japanese firms, will probably increase (Jonash and Womack 1985; Sugiyama and Fujimoto 2000; Humphrey et al. 2000). Toyota, for example, the least international of the Japanese auto firms, moved to expand its ASEAN presence in the early 1980s as a response to declining demand and rising import barriers in Western auto markets (Tarmidi 2004).

More specific features of Japanese corporate strategies reinforce the impact of these long-term market considerations. Japanese auto firms incorporate long-range market concerns into long-term investment decisions. An extensive comparative study of Japanese and German auto firms operating in Indonesian automotive clusters concluded that the Japanese time frame for returns on investment is often ten to 20 years, much longer than those of German counterparts (Doner 1991; Aswicahyono et al. 1999; Rasiah 2005; Rasiah et al. 2008). The Japanese emphasis is on long-term market share and not short-term profits, and this long-term investment perspective is strengthened by the view that investment in the individual ASEAN countries is necessary to create footholds in the emerging clusters in the Asia-Pacific region (Sugiyama and Fujimoto 2000; Yoshimatsu 2002; ADB 2005; Ozawa 2005; UNCTAD 2007).

Unlike push factors, pull factors normally work to assist a company in deciding on one location among several alternative sites. For both the Japanese and other Asian governments, captains of industry and government have been collaborating closely on sustaining the development and progress of the auto industry for both parties (Rasiah et al. 2008). However, from the host country point of view, in making this assertion, it is argued that Indonesia has both a bureaucratic authoritarian and a centreless state. The cooperation between government–business networks has been maintained slowly (Hatch and Yamamura 1996). It is referred to as a 'bureaucratic-industrial complex'; while Indonesian's politicians have the final say on redistribution policies affecting special interests, bureaucrats more often than not get their way on larger or longer-term issues perceived to be in the national interest, such as foreign policy and economic planning (FAIR 1989; Doner 1991; ADB 2005).

This contrasts with the Japanese point of view, where government has penetrated business and business has penetrated government through a process Richard Samuels (1987) calls 'reciprocal consent'. In other words, Japan's government–business network is a mutually reinforcing alliance of partnership that is capable of strong, decisive action so long as it keeps to the established, conservative policy line (Tokunaga 1993; Kodama and Kiba 1994; Hatch and Yamamura 1996). Furthermore, the Japanese government–business network has toed the line carefully in Asia, particularly in Southeast Asia, which has long been identified as critical to Japan's national security. It has tried to cultivate close relations with elites in the region, aimed at securing political and social stability, as well as the liberal trade and investment policies that are vital to Japanese capital (Sekiguchi 1983; Miyakawa 1991; Kiba and Kodama 1991). In fact, the 'trinity programme' has been initiated, known as 'comprehensive economic cooperation' with Southeast Asia.

At first, Japan's MNE automotive economic cooperation policy in Southeast Asian automotive clusters was based on the need to secure a steady supply of raw materials and a low-cost production base.

However, by the mid-1980s, the ground beneath that policy shifted when dramatic appreciation of the yen undermined the international competitiveness of virtually all manufacturing enterprises that export from Japan (Hideki 1988; Mai 1991; Borrus 1992; Kodama and Kiba 1994; Ozawa 2005).

As a result, the Japanese automotive industry began to see the region as an extension of its home base (Tokunaga 1993; Hatch and Yamamura 1996; Terry 2002). This led to the government-business network promoting a new vision of Southeast Asian automotive clusters as integral parts of a Greater Japan, with critically important links in an expanded Japanese production and exporting alliance (Hideki 1988; Kayoko 1991; Borrus 1992; ADB 2005). Japanese government–business believes the globalisation of economic activity has made it impossible to push ahead with economic development within the limited framework of a country defined by strict national boundaries, particularly in the Asia-Pacific region as 'one large economic zone and centre of the growth' (FAIR 1989; Mathew and Cho 2000; UNCTAD 2007).

Vertical and horizontal networks in the automotive industry

In shorter product development rapid technology diffusion, quicker product obsolescence, and the proliferation of quality producers worldwide have forced automotive firms to realise that they cannot excel in every aspect of their business system (Nonaka and Takeuchi 1995; Nonaka et al. 2000; Amin and Cohendet 2004). In view of that, according to the MNEs' vertical and horizontal networks, economic activity is organised through cooperative relationships among five partners in a business network. Key suppliers are expected to give near or total exclusivity to the flagship firm (i.e. an automotive MNE). In return, they benefit from increased volume and a greater portion of the value added to the product (Rugman and D'Cruz 2002).

Additionally, the theory of business networks in the field of business strategy and international management explains how a framework of vertical and horizontal networks addresses current managerial concerns linking the role of MNEs to issues in business policy and organisational learning (Rugman and D'Cruz 2002). The five partners consist of automotive MNEs (regarded as the flagship firm), key suppliers, key customers, competitors, and other stake holders (non-traded service sectors, government, social services, and education).

In the same way, relationships with key customers are marked by collaboration and sharing of information and resources. Furthermore, competitors can enter into more cooperative relationships with the automotive MNEs through market sharing arrangements, joint R&D projects, shared training ventures and facilities and supplier development programmes (Henderson 2002; Pedersen et al. 2003). As a result, collaboration within the business networks allows partners to accelerate organisational learning through access to the resources and expertise of other organisations (Rugman and D'Cruz 2002).

Like most complex business strategies, building a network is best accomplished one step at a time rather than in a few sweeping moves. Therefore, rather than invest vast effort in an elaborate strategic plan complete with details of the many steps needed to implement an effective network strategy, the automotive MNEs (positioned in the middle as a 'flagship firm') should instead undertake many small initiatives and build network linkages with suppliers, customers, and competitors wherever appropriate (Lundan 2002; Gallaud and Torre 2004).

In the Japanese automotive business network, MNEs with global scope must forge cooperative relationships with their partners in their business systems (Ichiro 1991; Florida et al. 1998; Rutherford 2000; Ozawa 2005; Fujimoto 2007). Flagships firms in the Japanese automotive industry continue to compete with other each other. The five leading Japanese manufacturers – Toyota, Honda, Nissan, Mazda and Mitsubishi – compete aggressively with each other for global sales and market share. Instead of developing and using market power to gain competitive advantage, the firms must learn to collaborate in cooperative relationships aimed at enhancing their mutual competitiveness. Subsequently, in a mature key supplier relationship, a key supplier can regain influence on the strategic direction of the business network. Tiered networks especially in the automobile sector have significant importance in the integrated production system (Fujimoto 2007).

As the Japanese networks discovered many years ago, product and process development have to be closely linked due to JIT (Just in Time) and the *kanban* systems (Miyakawa 1991; Charles and Li 1993; Dyer and Nobeoka 2000). The automotive flagships and their Tier 1 suppliers supply manufacturing components, and their development capabilities are important to the process development, which can only take place in close proximity to operations (Larsson 1999; Boschma and Wenting 2004; Rasiah 2005; Rasiah et al. 2008). Therefore, forging and managing vertical relationships with suppliers is necessary for the mutual interest of the partnership to replace the self-interest of each partner (Mikler 2007; Rasiah et al. 2008; JAMAI 2008).

Subsequently, as operations activities in automotive components supply increasingly migrate to Tier 2 suppliers, it is conceivable that development activity should follow. In such mature networks, it is likely that the following will occur (Chen 1996; Fujimoto 2007):

1 Flagship firms will relinquish much or all of their technological leadership role.
2 Tier 1 suppliers will become systems integrators and assemblers who will focus their resources on understanding and anticipating the systems requirements of the original equipment manufacturers (OEMs), in addition to managing a network of relationships with Tier 2 suppliers.
3 Tier 2 suppliers will learn to play a leadership role in those areas of technological development that require a combination of product and process development capability.

FDI in the Japanese automotive industry and its supply chain networks in Southeast Asia

It has been argued that the large automotive multinationals would tend towards increasingly internationalised production networks (Dicken 1988; West 2000; Amin 2002; Narula 2003; Ozawa 2005). Each company would produce a pool of strategic components (engine, suspension system, gearbox), from plants established anywhere in the world, to produce parts at the most efficient scale possible (Vernon 1996; Fujita *et al.* 1999; Henderson 2002; Lundan 2002). Other components would be bought in from outside suppliers at a low price because of the quantities required (Rutherford 2000; Busser and Sadoi 2004; Andersen and Christensen 2005).

Because of the uniform basic design of car industry, competition would be based on price, and thus production technology and manufacturing location would be characterised by very large economies of scale at low labour cost locations (Klepper 2004; Pries and Schweer 2004; Fujimoto 2007). In order to keep costs and prices down, a geographical shift of production from the major markets in developed countries to a cheaper labour cost location in newly-industrialising countries was envisaged (Dicken 1988, 2003; Fujita and Krugman 2004; Coe *et al.* 2007).

Some scholars (Womack *et al.* 1990; Kenney and Florida 1993; Florida *et al.* 1998; Rutherford 2000) predict that such an economy will emerge as globalisation and direct competition between different productions systems lead to the survival of the most efficient economies, particularly in the context of Japanese automotive MNEs (Ichiro 1991; Tokunaga 1993; Terry 2002; Narula 2003). Under these circumstances, inward automotive FDI serves as a transmission vehicle for the best practices of the investing regions (Busser and Sadoi 2004; Mikler 2007).

Additionally, as a result of the Asian expansion through Japanese FDI, the potential for auto production by developing countries gave a similar impression when ASEAN-4 clusters (Thailand, Indonesia, Malaysia and the Philippines) initiated their automotive efforts in the late 1960s and early 1970s (Hideki 1988; Doner 1991; Han 1994; Yoshimatsu 2002; Narula 2003; Rasiah 2005).

In line with that, Japanese dominance of ASEAN-4 auto markets has been accompanied by an extensive growth of Japanese manufacture, assembly, and parts production in the region (Hatch and Yamamura 1996). For that reason, the Japanese are at the leading edge of technological and competitive developments in the industry. ASEAN-4 clusters have been a major overseas focus for Japanese rivalry, encouraging the automotive product cycle further (Yoshimatsu 2002). In view of that, as technological factors affect the ability to exploit this potential, the exploration of MNEs in the automotive industry is an expanded benefit expected by the host country (Rasiah 2005; Rasiah *et al.* 2008).

In the case of Japanese automotive makers, global-local manufacturing has been an important strategy for expansion (Tarmidi 2004; Fujimoto 2007; Honda 2004). Take the case of Toyota and Honda, which have built up their global-local parts sourcing alongside the opening of new plants in selected host developing countries, with complex, high levels of parts-localisation, which is essential for efficient and timely manufacture (Dicken 1988, 1992, 2003; Doner 1991).

Additionally, ASEAN-4 clusters are perhaps the most suitable in the world for Japanese expansion in the automotive industry (Aswicahyono *et al.* 1999; Yoshimatsu 2002; Damuri *et al.* 2006). Besides the fact that the region is right next to door to Japan, it is still developing. Wage levels are rising quickly, but remain low compared to those in the developed world. What is more, ASEAN-4 clusters are filled with 'developmentalist-minded' governments that are eager to have their economies raised not only by Japanese capital and technology, but also by Japanese guidance on government intervention and industrial organisation (Dicken 1988, 2003; Hatch and Yamamura 1996; Borrus 1992).

As a result, Indonesian automotive clusters now seek to regionalise the dense web of mutually reinforcing ties between government and business, business and business, and management and labour (Tarmidi 2004; Wee 2005; Rasiah 2005). In other words, cooperation is the principle that informs relations between Japanese automotive makers and the host country. The long-term contractual relation and integration in *keiretsu* between Japan and the host country is a long-term process because of the complexity in contracts which might impose high costs in transforming, monitoring, and enforcement (Lindsey 1985; Doner 1991; Chen 1996; Liu and Dicken 2006).

Although the neoclassical economist view is that Japan is really 'doing nothing extraordinary' at all in Asia and ASEAN (Aoki 1988; Miyakawa 1991; Mair 1994; Ernst 2000; Yoshimatsu 2002), this research intends to demonstrate that Japan is not only plugging into the region's economic energy but also transforming and promoting the host country in technology-based production alliances in the automotive sector.

In view of that, all these investments and tie-ups are nothing less than the regionalisation of Japan's vertical or supply *keiretsu*. In this way, Japanese high-technology and high-volume MNEs have been able to replicate the core of their quasi-integrated production regimes to reduce transaction costs and, if regionalised, might come to generate efficiencies for many years (Coe *et al.* 2004, 2007; Beck 2005; Gereffi *et al.* 2005).

These are difficult but not insurmountable problems. What truly stunts the growth of local suppliers is the fact that Japanese MNEs in this region are building a tight network of dedicated suppliers from Japan, but a far looser, or wider, network of domestically owned suppliers (Dicken 1988; Yeung 2000; Coe *et al.* 2004). In other words, they are employing what some call 'market sharing agreements' and others call 'multiple sourcing' – a practice whereby large assembly firms purchase the same or similar product from different suppliers at different times (Dicken 2000; Coe *et al.* 2004, 2007; Dicken 2007).

Market sharing agreements, which MNEs thrust upon their suppliers and subcontractors, act as a deterrent to industrial upgrading. The quantities ordered from each supplier are enough for minimum production runs but insufficient for higher volumes where economies of scale can be derived through better technology, rationalised production lines, and improved management techniques (Dicken and Malmberg 2001; Dicken 2007). Deliberate sourcing policies such as those pursued by Japanese companies provide no incentives for industrial deepening or upgrading by local firms.

Accordingly, as more and more Japanese subcontractors respond to home and host government incentives by investing capital or licensing technology in automotive clusters in Southeast Asia, native suppliers seems to get less and less action, resulting in protests to government officials throughout the region (Okada 1983; Hatch and Yamamura 1996; Ernst and Kim 2002).

Protests have not paid off. To get a piece of the action, local suppliers often must swallow hard and relinquish control to Japanese managers by entering into a joint venture or technical tie-up. However, try as they might, local business people cannot always convince Japanese businesses to tie the knot (Okada 1983; Wickens 1995; Sugiyama and Fujimoto 2000). To some extent, Japanese automakers do offer reasons to explain their strong preference for Japanese transplants rather than native suppliers. For instance, local suppliers cannot or will not keep up with the delivery schedule, causing them to shut down assembly lines as they wait for shipments of needed inputs. This is obviously no way to run a JIT production system. Still others complain loudly that local suppliers quite often fail to meet their minimum standards for quality (West 2000; Liker 2004a, 2004b; Liker and Meier 2005).

It is difficult, if not impossible, for local suppliers to keep pace with Japanese assemblers and non-Japanese joint firms that are making what have been described as day-to-day innovations, or frequent changes in production or process technology originating in Japan (Andersen and Christensen 2005). As a consequence, rather than just wringing their hands, several Japanese MNEs are trying to help local suppliers meet their expectations (Tarmidi 2004; Honda 2004).

In the quest for efficient supply networks, Japanese automakers in Asia are doing something that American manufacturers would never dream of doing (Hatch and Yamamura 1996; Aswicahyono and Anas 2000). They are teaming up to form what could be considered 'super *keiretsu*'. For instance, Toyota and Daihatsu have agreed to use some common components for the family wagon car for the Asian market. In addition, Suzuki and Mitsubishi Motors also agreed to produce joint truck programmes. Likewise in Thailand, Toyota, Nissan and Isuzu have begun to collaborate on the production of cylinder blocks for diesel engines. This cooperation was designed partly to satisfy the demands of the ASEAN market and partly to maintain Japanese domination of the local market (Aswicahyono and Anas 2000; Rasiah 2005; Damuri *et al.* 2006).

For non-Japanese firms, this means that local suppliers always face an uphill battle in trying to establish credibility. Thus, establishing a business relationship with large Japanese MNEs remains tough.

However, it is argued that Japanese developmentalism through Japanese FDI in the automotive industry in Southeast Asia's clusters, and particularly in Indonesia, has generated benefits to this region (Doner 1991; ADB 2005; UNCTAD

56 Production and knowledge transfer

2007). Under this 'developmentalism', innovating manufacturers in the automotive industry rapidly increased their productive capacities, turned to exports, and began achieving dynamic technological efficiency (Sugiyama and Fujimoto 2000). Along with the largest firms created and maintained by *keiretsu* networks, the quasi-integration of subordinate firms by dominant firms increased the international competitiveness of Japanese high-tech industries.

For Indonesian auto firms, the benefits of developmentalism via quasi-integration are many, particularly in the early stage of network formation (Odaka *et al.* 1988; MOTIRI 2005). That is when these firms receive invaluable infusions of capital, technology, and managerial guidance from the Japanese government–business network. Although there is still unequal cooperation, the production alliance now emerging in Asia is in its early stages; the benefit it is producing for Asian economic growth still exceeds the costs it imposes (Yeung 2000; Schmitz 2000; Terry 2002; Wolf 2004).

A risk for the local auto firms in ASEAN-4 clusters is that they may become stuck in the mechanisms of Japanese automotive *keiretsu* (Soesastro 1989; Doner 1991; Sugiyama and Fujimoto 2000), which has long-term implications that they will become heavily dependent partners in vertical networks (i.e. supply chain) without having indigenous power to initiate a localisation project. It has to do with asset specificity; since most of the physical and human capital of the subordinate firms is dedicated to maintaining its relationship with dominant parent companies, the subordinate firms are exposed to constant demands regarding price, quality, and time. The parent companies, in other words, are able to squeeze the subordinate firm as it strives to increase its profitability and international competitiveness. The subordinate firms often have little choice but to bow to pressure if they wish to maintain the value of their assets and continue to benefit from an ongoing relationship with the dominant partners (Henderson 2002; Hess and Yeung 2006). However, in the case of Toyota and Honda in Indonesia, the local transplants have shown an indigenous initiative and they have been able to initiate local projects for cars and motorcycles, developed by Indonesian engineers with the help of the parent company in the knowledge transfer process. These findings are explained in Chapters 7 and 8.

In general, the Japanese automotive *keiretsu* scenario is reflective of what has happened in the case of indigenous auto part companies that are not part of Japanese automakers' first or second tiers (West 2000; Rugman and D'Cruz 2002; Fujimoto 2007). The unequal bargaining power among indigenous auto part firms is a problem in the tight auto industry. If they do not belong to the Japanese *keiretsu*, it is hard to penetrate a crowded market which is under Japanese control (Doner 1991; Chen 1996; Ozawa 2005; Fujimoto 2007).

It must be acknowledged that Japan is moving too quickly to stake its claim on Asia; Pax-Niponica or East Asian mutual benefit zones (Hatch and Yamamura 1996; Ozawa 2005) are dominated by Japan. In fact, despite the lure of Japanese capital and technology, and despite the attractiveness of the Japanese model of economic development, there are a growing number of signs that other Asians are not comfortable with their subordinate role in the production alliance now taking shape in the region. Japan is supposed to be the engine of growth, not the mechanism for a path to dependency (Terry 2002; Ozawa 2005).

With this in mind, there are some concerns to be taken into consideration by Asian governments to ensure their economies do not become captive members of a Japanese production alliance. MNEs under pressure from globalisation in the Asian region must do more to increase their own technological capacities (UNCTAD 2007; IMF 2007). This means investing wisely in education, training, and creating stronger links between public research facilities, particularly universities and private industry.

This initiative is stated in FDI agreements: when the MNEs transfer their technology, it should benefit the local community in comprehensive ways (economy, social-cultural, education, technology) (Liu and Dicken 2006). Subsequently, when governments have adopted measures designed to promote supporting industries, they have often ended up assisting foreign MNEs to establish domestic facilities rather than domestically owned supply firms (Doner 1991; Markusen 1999; ADB 2005; UNCTAD 2007; IMF 2007).

Nevertheless, it is difficult to draw a firm conclusion about whether a MNE location will be beneficial or malignant for the regional economy without taking into account all the costs and benefits that arise from attracting such investment (Krugman and Venables 1995; Metcalfe and Ramlogan 2006; Mikler 2007). Since the inward overseas investments to a region are an integration of widely diverse economic interests, it is only by focusing on the complexity of products and processes of investment flow that an unambiguous understanding of the investment role in the regional economy can be gained (Han 1994; Doner 1991; UNCTAD 2007; IMF 2007).

Accordingly, the negative aspect of Japanese FDI automotive investments in Southeast Asian clusters have resulted in the increased external control of a region, and, thus, a branch plant economy or a loss of structural autonomy (Soesastro 1989; Doner 1991; Okada 1983; Sugiyama and Fujimoto 2000), for instance, the reduction of local linkage, diminishing R&D activities and skilled labour employment and prevention of local initiatives. In a similar vein, the vulnerability of the host country economy to international demand and supply conditions results from MNEs activities based on global conditions (Dicken 2000; Henderson 2002; Hess and Yeung 2006).

Conclusion

Knowledge transfer in the automotive sector in technology and productivity is neither automatic nor easy. This is because it depends on investment in tangible capital, as well as intangible capital in the form of education and training – at least in the industrially advanced countries – of business expenditures on R&D and related activities (Fagerbergh 1993; Krugman and Venables 1995; Dicken 2003; Coe et al. 2007). These factors explain why some automotive clusters in developing countries have been successful in reducing the technological and productivity gap, while others have not.

Knowledge production through knowledge transfer in this sector is no longer perceived as a linear process, but instead as the result of the complex interaction between numerous actors and institutions which include codified and tacit knowledge (Nonaka *et al.* 2000; West 2000; Nooteboom 2004; Liu and Dicken 2006). However, firms, organisations and institutions, as well as their interactions, differ substantially between countries. This implies that policy responses to systemic imperfections will be country-specific.

Nevertheless, knowledge transfer from the Japanese automotive industry in Southeast Asia's clusters has been acknowledged as a positive skill formation and technology. The Japanese automotive MNEs have gradually transformed this region from a low-wage and intensive labour industry into the Japanese automotive export based region. The rapid expansion of production, the integration of local production into a global production network and the integrated global supply networks by Japanese automotive MNEs have been energising the ASEAN-4 automotive clusters up to the present time. Despite the difficulties, it is clear that the globalisation of Japanese automotive MNEs have influenced considerable changes in the automotive industry in the ASEAN-4 automotive clusters that host Japanese investments in this sector.

In this chapter, I have outlined the theoretical justifications for taking knowledge transfer elements to the vertical and horizontal networks of Japanese automotive MNEs and the pragmatic reasons why the automotive MNEs are important players in studying the automotive cluster. I conclude that the MNE automotive industry can be beneficial to the host country if properly applied. Therefore, my empirical work goes on to focus on the process of knowledge transfer of MNEs in the Indonesian automotive clusters and examines the way in which cluster policy has been beneficial for the growth of the Indonesian automotive clusters. The emphasis is on how the role of MNEs makes an impact on the host country in developing the production system and enhancing the integrated global production network.

4 Global and national environments
The macroeconomic context in Indonesia

The global and national environment, as described in Chapter 1, is explained further in this chapter, which deals with the macroeconomic background of Indonesian industrial policy and the Indonesian automotive industry; and in Chapter 6, which deals with manufacturing conditions in the Indonesian automotive cluster.

Chapter 4 is a prelude section to the discussion of the automotive cluster in Chapter 6 and an outline of the case studies in Chapters 7 to 9. Chapter 4 illustrates the Indonesian macroeconomic situation in the aftermath of the financial crisis in the late 1990s. In addition, this chapter highlights Indonesian industrial policy on the automotive industry and how this policy links to the automotive clusters, issues which underpin the research questions and research background of Chapter 1. Moreover, Chapter 4 traces the historical trajectory of the Indonesian automotive industry before and after the Asian financial crisis.

This chapter also gives an overview of historical explanations for cluster implementation in Indonesia, as it occurred in Indonesian small- and medium-sized enterprises (SMEs) before the advent of hi-tech industries such as the automotive industry. The chapter also underlines the significance of industrial estates that were the origin of industrial clusters in the Java region. This chapter also explores the current progress of industrial estates and their significance for developing cluster strategy in the Indonesian industrial context.

In addition, Chapter 4 provides a framework for the understanding of Indonesian cluster strategy and automotive cluster which is explained in further detail in Chapter 6, and serves as a useful preface to the case studies of Toyota and Honda in Chapter 7 and Chapter 8 respectively.

The Indonesian macroeconomy after the Asian financial crisis

The global environment has influenced the Indonesian macro- and microeconomy, particularly the industry and trade sector which still depends on multinational enterprises (MNEs). Therefore, as indicated in the conceptual framework in Chapter 1, the impact of global environment is influencing the Indonesian government to improvise its industrial policy, in this case, the aftermath of the Asian financial crisis. Up to 2007, Indonesia's economic growth performance improved following a slow recovery from the Asian economic and financial crisis between 1998 and 2000. The crisis was compounded in Indonesia by an ensuing political crisis that resulted in the end of the government led by former President Suharto (World Bank 2005; ADB 2005). Successive governments have prioritised economic stability during the challenging years since 1998.

Despite the severe difficulties that had accumulated from the 1998 period, the Indonesian economy managed to achieve a steadily increasing gross domestic product (GDP) growth in the five years between 2000 and 2005, approaching 5.5 per cent in 2005. This was a great improvement from 1998, when GDP contracted by 1.5 per cent (BPS 2004, 2005).

Looking forward, the current government is seeking to restore economic confidence through a development programme that prioritises a robust industry and trade policy alongside a science and innovation strategy for 2025 (MOTIRI 2005). Moreover, in order to maintain and enhance the investment climate and growth, the government introduced a new package of investment policies regulated under the Presidential Instructions (Inpres) no. 3/2006. These new policies cover customs and excise, taxation, labour and SMEs. As a result, it is expected that these will create a friendly investment climate and help boost MNEs through foreign direct investment (FDI) channels.

Additionally, the implementation of the Asian Free Trade Area (AFTA) since 2002 and membership of the World Trade Organization (WTO) has triggered another challenge to Indonesia, that is, its ability to compete with other neighbouring countries in manufacturing and the marketing of its products (ADB 2005). For that reason, the government's main strategy for the future development of the manufacturing sector is to boost manufacturing investment in industries which have competitive and comparative advantages. In the mean time, the government continuously strives to improve political, social, and economic conditions in order to create a competitive investment climate that will boost investment value in all sectors of industry.

Indonesia has now fully recovered from the Asian financial crisis of the late 1990s (OECD 2008). However, the country's post-crisis adjustment has been slower than its regional peers (World Bank 2007; OECD 2008).

Investment suffered the sharpest decline at the time of the crisis, a development that can be attributed to a large extent to a reversal in FDI inflows. From the supply side, the turnaround in manufacturing value added has also been slower in Indonesia than in comparable countries, although it had recovered to pre-crisis levels by 2000 (OECD 2008). This lack of dynamism in manufacturing growth after the crisis poses challenges for the future. Moreover, inflation has been higher in Indonesia since the crisis than for its regional peers. This is due in part to the large nominal depreciation of the rupiah during the crisis (World Bank 2007). Also, Indonesia's export growth has been the lowest among the crisis-hit countries, especially in manufactured goods. The contraction in exports was the sharpest in the region in the wake of the crisis, although growth has picked up in recent years.

Manufacturing investment in Indonesia

Indonesia's manufacturing sector was formerly the most dynamic sector of the economy. Widely known as one of the ASEAN Tigers, it had achieved a persistently high growth of around 11 per cent per annum (p.a.) during the 25 years preceding the Asian economic and financial crisis (World Bank 2005, 2007). The growth rate of the manufacturing sector, however, was most affected by the economic crisis. Despite the increase in overall GDP growth, this sector still does not seem to show the upward trend in job creation required by the country to sustain real growth.

The manufacturing sector contributed about 28 per cent of Indonesia's total GDP in 2001–2005, which was a slight decrease from 29.7 per cent in 2000 (ADB 2005). However, the value of Indonesian manufacturing investment in 2005 slightly dropped to −3 per cent of its 2001 value. Furthermore, the portion of the total Indonesian workforce employed in the manufacturing sector decreased from 13.3 per cent in 2002 to 12.3 per cent in 2005. The decline was apparently attributed to the slight decrease in manufacturing investment during the same period. The Indonesian government has been trying its best to restore economic confidence through development programmes that prioritise specific sectors and aspects as formulated in the new investment policies package (MOSR 2002). It is expected that with a stimulating investment climate, along with improved political and macroeconomic fundamentals, Indonesia would then be able to attract many investors and continue its economic recovery.

The Japanese automotive MNEs are recognised as having a key role to play in restoring strong economic growth, improving knowledge transfer, and creating new employment opportunities. While the Indonesian economy still suffers from many implementation and law enforcement hurdles, it is believed that in the long term, with the implementation of pro-market economic and financial policies, the country has the potential to achieve strong sustained growth over the coming years.

Moreover, Japanese automotive expansion is an important source of financial capital accumulation in Indonesia and the ASEAN region. In fact, Indonesian net FDI inflows have recovered in recent years, following the financial crisis of 1997–1998, although it is relatively slow compared to Singapore, Thailand, and Malaysia. Furthermore, Southeast Asia was among the most attractive destinations for automotive MNEs outside the OECD area in the first half of the 1990s, a situation that changed radically following the financial crisis. Only recently have FDI inflows recovered to their pre-crisis levels in most countries (UNCTAD 2007).

Indonesian investment, both foreign and domestic, is evenly concentrated across sectors, provinces and residency of foreign investors (ADB 2005). While domestic investment has focused predominantly on labour-intensive sectors such as paper, food processing, agriculture, and construction, there has been considerable MNE involvement in more capital-technological intensive activities, such as the transport/automotive sector, communications, and the chemical and pharmaceutical industries (ADB 2005; MOTIRI 2005). Along with that, the geographical distribution of investment is also concentrated in specific areas; foreign investment tends to favour locations in Java (80 per cent of total MNEs), whereas domestic firms invest in both Sumatra and Java.

Accordingly, efforts have been made to promote investment opportunities. The authorities have the intention of converting the Investment Co-ordinating Board, created in 1973 as a screening and authorising agency for foreign investment, into a fully-fledged investment promotion agency (Wee 2000). Additionally, they intend to strengthen the Board to improve coordination among the various government agencies involved in investment regulations, are working towards reducing the number of procedures needed for approval of new investments, and intend to cut back the approval period to one month from the current 105 days (Wee 2000; Soesastro 2003).

Initiatives are underway at local government level to facilitate the issue of business licences. Several local governments are setting up business licensing centres as a means of dealing with the uncertainty associated with the proliferation of local business regulations (Soesastro 2003; Perdana and Friawan 2007). As in other decentralised countries, the licensing process involves many procedures at different government levels. Verification of compliance with zoning rules and health and safety standards, as well as the issue of tax registration documentation and product- or activity-specific licences, require involvement not only of the national and local governments, but also of local business associations. Therefore, one-stop shops (OSSs) have been set up to consolidate the processing of business licences issued by separate bodies. The Minister of Home Affairs has recently issued general guidelines on how to establish these regional OSSs.[1]

Despite the recent efforts to liberalise Indonesia's investment policy regime, the remaining restrictions are relatively burdensome by international comparisons. On the basis of a 2008 OECD report, Indonesia is stricter than most countries in the OECD area, except Australia, Iceland, and Mexico. This implies that those latter three countries impose more restrictions of foreign investment than Indonesia. Nevertheless, Indonesia fares better than the BRICS (Brazil, Russia, India and China) group of countries, except Brazil, and South Africa, suggesting that it is relatively well-placed in relation to other major emerging market economies, which are among the most attractive destinations for MNE investment outside the OECD area (OECD 2008; UNCTAD 2007).

Subsequently, consideration could be given to a further relaxation of remaining barriers to foreign investment. In particular, foreign ownership constraints in selected sectors may have a detrimental impact on the quality of FDI. While intended to encourage technology transfers from MNEs, this requirement may in fact discourage such transfers, because foreign investors may be worried about losing intangible assets to local partners. Therefore, policy efforts to boost human capital accumulation could have the additional pay-off of removing an obstacle to foreign investment in knowledge-intensive sectors (MOTIRI 2005).

Before the 1990s, technology transfers through MNEs were limited in Indonesia due to capacity constraints to absorb foreign technologies, including skills shortages. These weaknesses will need to be tackled if Indonesia is to diversify its exports from primary and labour-intensive goods towards manufactured goods with higher value added. Therefore, the automotive sector aims to be the main prime mover for the Indonesian Vision 2025 of a more advanced industrial nation (MOSR 2002; MOTIRI 2005) in order to enhance knowledge and capability in Indonesia. Nevertheless, at the same time, there need to be efforts to develop infrastructure, particularly transport and telecommunication, with regards to the geographical condition of the Indonesian archipelago. This would enhance Indonesia's attractiveness to foreign investment in an environment of heightened competition for automotive MNEs, specifically among regional peers (UNCTAD 2007; OECD 2008).

Indonesian industrial policy in the automotive sector

The Indonesian automotive industry and ASEAN automotive clusters

The financial crisis of 1997 that started in Thailand and soon spread to most other ASEAN car base production countries had serious repercussions for the automotive industry in this region (Mikler 2007). Sales declined and production was put back. However, Japanese dominance of the automotive industry in Southeast Asia remained firm, although American and European makers gained some market share in the period following the Asian crisis (Sato 2001; Busser and Sadoi 2004; Rasiah 2005; Wee 2005).

The crisis deepened Japanese involvement in the ASEAN automotive sector because Japanese investors came up with additional investments in existing joint ventures to save these endangered enterprises.

The automotive industry in Indonesia and Southeast Asia has become increasingly integrated across international boundaries. Indonesia has been working hard to get back on the right track after being devastated by the economic crisis of the late 1990s (Gaikindo 2007). Therefore, an open trade policy by the Indonesian government ensures that there is major industry rationalisation in this sector. For instance, an open foreign investment regime combined with an effective industrial extension programme, and measures to promote technology transfer from foreign to local parties, will facilitate continuous improvement in supply-side capacities.

Accordingly, the recovery of the Indonesian automotive sector is due to the government's strong commitment to promote an encouraging business climate through automotive policy and industry. The government of Indonesia has been improving its policies for the automotive sector, to support the implementation of the regional and multilateral arrangements in effect. As a result, the industry could look forward to becoming one of Indonesia's prime movers in manufacturing, with much-improved efficiency and a growth path.

The Indonesian cluster policy was perceived by the automotive industry as a step towards encouraging networks among the Japanese FDI and its global-local *keiretsu* (MOTIRI 2005; Tarmidi 2004; Wee 2005). This policy has been in place since the 1990s but in the aftermath of the financial crisis the government decided to push the private sector to be involved in maintaining industrial clusters across Indonesia (MOSR 2002). This is partly due to the Japanese *keiretsu* joining the development of clusters in Indonesia. Specifically within the automotive industry, this cluster policy has been supporting production and marketing activities. Drawing on the Japanese Just in Time (JIT) production system, the cluster policy has supported the growth of MNEs in the automotive sector, and production and marketing are back on track after the economic downturn in the late 1990s (Takii and Ramstetter 2005).

The automotive industry in Indonesia is now seeking to regionalise the dense web of mutually reinforcing ties between government and business, business and business, and management and labour. In other words, cooperation is the principle that informs Japanese automotive makers' relationships with the host country. The contractual relation and integration in *keiretsu* between Japan and the host country is a long-term one because of the complexity of contracts, which might impose a high cost in transforming, monitoring, and enforcement (Lindsey 1985; Doner 1991; Chen 1996).

Moreover, in the ASEAN region, Thailand, Indonesia and the Philippines had already attempted to develop automotive industries. Within the greater industrial policy framework, the automotive industry was part of a broader import substitution developmental strategy (Busser and Sadoi 2004). Under the import substitution policies, complete knocked-down (CKD) automotive production increased in ASEAN-4 automotive clusters.

Nevertheless, the ASEAN governments soon realised that the benefits of this industry for their domestic economies were limited because the industry was still dependent on technology supplied by Japan (Tarmidi 2004; Rasiah 2005).

Accordingly, to raise a strong class of local suppliers and to push for technology transfer, most ASEAN countries developed so-called local content policies. The basic idea was to force automotive manufacturers to purchase locally both parts and components. The Japanese investors responded positively to this new policy (Ozawa 2005). As Japanese automotive makers increased their production volumes overseas, Japanese parts suppliers started to establish overseas production sites near their Japanese customers at overseas locations. As a result, investments by Japanese final-assembly automotive makers grew quickly along with part suppliers' investment. Thus, the role of local enterprises as suppliers to Japanese final assemblers or first-tier suppliers has grown.

Historical trajectory of the Indonesian automotive industry

The automotive industry in Indonesia has existed since 1927 when General Motors (GM) first built a car plant in Jakarta. However, it was not developed further and GM remained the only car assembler.

Cars and motorcycles were imported and until the 1960s the number was relatively small. Until 1969, the imports of automotives were not regulated and the industry consisted mainly of trading activities with very limited assembling operations (Tarmidi 2004). Nevertheless, assembling activities started developing in 1964 with semi-knocked-down components, which progressed into the assembling of knocked-down parts in 1971.

Subsequently, the Indonesian automotive industry became a significant but generally inefficient industry owing to a historically highly-distorted policy regime (Aswicahyono et al. 1999; Aswicahyono and Anas 2000). Nevertheless, the industry achieved some progress as indicated by growth and an increase in knowledge and technological capacities under continued support particularly from Japanese automotive makers (Gammeltoft and Aminullah 2004; Wee 2005; Rasiah et al. 2008).

Prior to the 1998 crisis, the Indonesian automotive industry recorded tremendous growth. Between 1994 and 1997, this sector experienced 20 per cent growth on average (Wee 2000). It accumulated Rp7.1 trillion in investment and employed 70,000 workers. Production of automotives in 1997 reached 390,000 units while several assemblers, especially of commercial vehicles, attained minimum efficiency (Aswicahyono et al. 1999).

The development of the industry has also created demand for components, for instance, local contents for minibus and pick-up vehicles were already 40 per cent, and 30 per cent for 5–24 tonne trucks. However, local content for sedans is limited to 10 per cent on average. Market fragmentation for sedans has hindered the efficient use of local components. Furthermore, while the automotive industry has increasingly used locally produced components, the components sector is still highly dependent on imported sub-components and raw materials (Okada 1983; Wee 2000; Rasiah 2005).

The economic crisis has exacerbated the inefficiency problem within the industry. While the industry faces an inefficiency problem, global trends may intensify competition within domestic and regional markets. However, the fact that Indonesia is an emerging market with rapidly expanding domestic demand implies that there is still an opportunity for developing the industry (Soesastro 2003; Wee 2005; Rasiah 2005; ADB 2006).

The development of the automotive industry in Indonesia can be explained in terms of import substitution based on industrialisation enjoying high levels of protection ranging from tariff to non-tariff barriers, to a relatively more market-oriented environment as the Indonesian economy has become more integrated to the world economy. Therefore, the historical trajectory of the development of Indonesian automotive policy can be explained as follows:

1 Pre-reform period

The Indonesian automotive industry was started when GM established an assembly plant in Jakarta in 1927. However, the development of the industry at the time was still insignificant as the activities involved were restricted to trading and simple assembling processes. Then, in 1950, GM was nationalised under Programme Banteng to build a national car industry (Tarmidi 2004).

Unfortunately, Programme Banteng came to an end in the early 1960s due to a shortage of foreign exchange. However, the ambition of having a national car industry re-emerged in the late 1960s when the new order government came into power in 1968. It placed an economic development programme as a priority in its national agenda in which a general industry strategy would develop an independent industrial sector.

In achieving its goal, it implemented an import submission policy. The automotive industry was considered one of the strategic industries (in conjunction with science and technology policy) to be developed as an import substitution industry for several reasons (Aswicahyono et al. 1999, Tarmidi 2004). First, it supplies the needs for transportation means; second, it not only contributes to economic growth and employment opportunities but also exposes the economy to knowledge and high-technology; third, it generates income for the government from import duties and taxes.

Additionally, the government had undertaken several important decisions in making the automotive industry one of the success stories of import substitution based on industry. Most notable was the import ban on completely built-up (CBU) cars in 1971 following the regulation requiring all investments in this sector to have domestic sole agencies in the form of joint ventures (Wee 2000). As a result, the sole agencies scheme was modified to become an assembly scheme in 1972. This scheme prohibited foreign automotive producers entering assembly and distribution activities since all investment in this sector had to be made in the form of domestic investment or joint ventures (Aswicahyono et al. 1999). These series of policies pushed automotive producers to assign a domestic company as a sole agent for importing cars in CKD forms, assembling the cars and then distributing the products.

Furthermore, the government protected the industry by implementing tariff and non-tariff barriers such as quota and local contents schemes (MOTIRI 2005). Consequently, many automotive companies and plants had to be created to produce several types of car. However, since these sole agents and assemblers originated from trading companies that generally have limited knowledge of car production and little motivation to develop the industry, there was no significant improvement attained in this industry until 1972.

2 Deletion and incentive programmes

In 1977 the government introduced the *Cinta Produk Indonesia* programme also known as the 'deletion programme'. The idea was to force domestic car producers to use domestically produced components and to support the growing industry. Accordingly, domestic producers were required to increase the use of domestically produced components, starting with the use of universal components, and moving towards more specific and targeted components. Under this programme, manufacturers who could not comply with the schedule would be sanctioned with 100 per cent import duty on the imported components that would otherwise have been domestically produced. It was expected that by the end of the programme, the automotive industry would be less dependent on imported components (MOTIRI 2005).

The goal of the deletion programme was never achieved (Tarmidi 2004). Several reasons were attributed to the failure of the programme namely: (1) low technological capability; (2) profits as a distributor were higher than those as a full manufacturer; (3) low production scale due to the very diversified brands and models produced prevented the industry from achieving economies of scale; and (4) foreign principals preferred to keep their sole agents as a distributor instead of a full manufacturer. As a result, only a small number of components such as lamps, accumulators, and tyres could be produced in large and economically efficient numbers.

In 1993 the deletion programme was abandoned and replaced instead by an incentive programme (MOTIRI 2005). This programme also aimed to promote domestically produced components. However, unlike the deletion programme, the emphasis was on giving an incentive to producers for using more domestic components (Tarmidi 2004). This incentive took the form of a lower import duty for imported components, sub-components, semi-finished parts and raw materials based on the degree of local content achieved. The higher the level of local content achieved, the lower the import duty that would be applied for the remaining components to be imported.

The incentive programme is based on a points system where each component (or group of components) is assigned certain points which are then used as a count of local content used by the manufacturer (Aswicahyono *et al.* 1999; MOTIRI 2005). The calculation of local content is relatively complicated, based on production process and weighted average of the type of main components and the type of the automotive.

Since its launch, the incentive system has gone through several revisions (MOTIRI 2005). The 1995 deregulation package granted zero import duty for the remaining components of commercial vehicles that had reached 40 per cent local value added and passenger cars that had reached 60 per cent local value added; 10 per cent import duty for the remaining components of vehicle production that achieved 20 to 40 per cent local value added; 15 per cent import duty for the remaining components that reached 20 to 30 per cent. In short, government policy was biased in favour of the development of commercial vehicles.

Nevertheless, despite this incentive provided to manufacturers, the development of the automotive industry was still disappointing (Tarmidi 2004). The local content of cars produced in Indonesia was relatively low. In 1995, local contents of sedans were only around 11 per cent while for commercial cars it was 40 per cent. As this deregulation package was implemented, it could be gauged that the government was moving in the direction of being more market-oriented in developing this industry.

3 National Car Project

A controversial government move in the beginning of 1996 made people question the government's intention of being more market-oriented (Aswicahyono et al. 1999; Wee 2000; Rasiah 2005). Former President Suharto instructed three economic ministers to coordinate steps towards the development of a national car industry.

This would mean cars would be produced within the country under a domestic brand, by a company entirely owned by Indonesians using entirely local components. Officially, the goal was not much different from previous policies. However, this policy was considered a necessary step in accelerating the process of developing a national car (Tarmidi 2004; Rasiah 2005). A special incentive was given to the establishment who would be given 'national car' status. Thus, the appointed company with this pioneer status would be required to achieve a local content based on a specific schedule (i.e. local content up from 20 per cent to 60 per cent by the end of the production process).

Subsequently, the government appointed PT Timor Putra National (TPN), owned by Tommy Suharto, the youngest son of former President Suharto, to be the sole carrier of national car status. The company was newly established with no track record and, even worse, without an automotive plant to manufacture the national car (Soesastro 2003). Furthermore, PT TPN planned to produce the Timor cars on a full manufacturing basis with technological and design assistance from KIA Motors Corporation of South Korea. Several plants would be built up to produce sedans and multi-purpose vehicles. At the first stage, TPN would ship KIA cars in CBU condition that would be sold in the domestic market at a lower price since it had national car facilities, although they were produced entirely in South Korea. Along with given status, the government also considered using state-owned banking to form a bank consortium to finance a PT TPN car assembling plant (Wee 2000).

Predictably, this policy attracted not only domestic criticism but also international criticism, specifically from car producing countries (Aswicahyono et al. 1999; Wee 2000). Japan, the USA and the European Community filed their complaints to the WTO concerning the national car programme and the exemption from import duties and luxury taxes on imports of 'national vehicles' and components and related measures. They also requested that the Dispute Settlement Body (DSB) at the WTO set up a panel examining the case.[2]

The panel recommended that the DSB request that Indonesia bring its measures into conformity with its obligations under the WTO agreement. On July 1998, the DSB–WTO ratified that the 1996 policy on the national car programme and the 1993 policy on tax incentives for local content achievement had to be abolished within 12 months, by 22 July 1999. It also referred to the elimination of local content requirements by 1 January 2000.

4 Commitment to the IMF to the present

At the end of 1997 when the economic crisis had intensified in Indonesia, the government of Indonesia asked for financial and economic assistance from the IMF (Perdana and Friawan 2007). In the agreed-upon assistance scheme, the government committed itself to the IMF's strict economic reform programme which covers actions in four major areas, namely: (1) the effort to restore the health of the financial sector; (2) fiscal policies; (3) monetary and exchange rate policies; and (4) structural adjustment, in the form of extension and deepening of a deregulation programme.

In the structural adjustment programme, there were several policy reforms that directly affected the automotive industry (Rasiah 2005; Wee 2005): (1) the commitment to reduce import tariffs; (2) barriers to export, including export taxes, would be reduced in stages; (3) following the commitments to the WTO, the local content programme for automotives, which had provided a high level of protection to the industry, would be eliminated by 2000. The government would also agree to whatever the WTO decided regarding the controversial National Car Project in 1996.[3]

As a result of the dispute with the WTO and as requested by the IMF, the Indonesian government decided to get the automotive sector back on track (MOTIRI 2005). Therefore, a new automotive policy was announced on 22 June 1999. It aimed at developing an efficient and globally competitive automotive sector. There are three aspects of the existing automotive policies that were revised in this package.

1   Elimination of the local content scheme. Under the new policy, imported cars and components are subject to import duties based on tariff lines, regardless of local contents achieved. The deadline for the elimination of the local content scheme is six months earlier than the deadline set by WTO.
2   Further reduction of import duties on imported cars and components. The largest reduction is in imported cars in a CBU condition, while import duties on CKD cars vary depending on their types. This is in line with Indonesian commitment under the ASEAN Free Trade Area (AFTA) and the Asia-Pacific Economic Cooperation (APEC).
3   Simplifications of procedures and qualification for importing cars. Vehicles in CKD forms can only be imported by assembling companies that have already been granted an import license while vehicles in CBU forms can be imported by general importers. At last, it helps promote the development of a component industry and export markets for automotive products.

The Indonesian motorcycle industry has not attracted much attention when compared to automobiles in the automotive regional meetings (AISI 2008). This is due to the fact that the motorcycle industry is ahead in its product life cycle at the global level. Although in its fast growth stage in large developing countries, not all APEC member countries have significant motorcycle industries. Within ASEAN, four countries (Indonesia, Malaysia, Thailand and the Philippines), have a motorcycle industry.

Looking at the appropriateness of the motorcycle as a cheap alternative to personal transportation for developing countries, it is indeed a very important automotive sector. While most if not all automobile manufacturing operations in Indonesia and ASEAN are suffering from a lack of economies of scale, motorcycle industries have enjoyed reasonably successful and efficient manufacturing operations, especially for the small utility motorcycle models such as the Cub Type, and small displacement sport types with engines below 150cc (Gunawan 2002; AISI 2008).

The most successful Indonesian manufacturer is currently producing around 4,500 units a day with a very high degree of local content achievement in components as well as engineering capability, particularly tooling and some manufacturing preparations (AISI 2008). Even some corporate R&D capabilities have been transferred to the ASEAN region (Honda 2004). Therefore, this transfer of manufacturing and engineering capability and of the R&D process allows motorcycle manufacturers to introduce models with a much faster turn-around time, more suitable to the taste of ASEAN customers, as well as cheaper products if compared to production and engineering base support in Japan. In fact, Thailand and Indonesia have become an important export base for Japanese motorcycle companies. This transfer of technology and manufacturing from Japan to the ASEAN-4 automotive clusters has been done step by step, and relatively smoothly since the late 1980s (Honda 2004; AISI 2008).

Due to its vast territory and large population, Indonesia is currently considered to be the third largest motorcycle market (after China and India) and the fourth largest producer in the world (after China, India and Japan). Production in 1997 reached an all time high of 1,852,906 units. However, due to the economic downturn since 1998, the market deteriorated significantly in 1998 and 1999 where production by PASMI (the Indonesian Association of Motor Cycle Assemblers and Producers) members was 517,914 and 584,114 units respectively (AISI 2008).

In 2000, Indonesia saw the arrival of large scale imports of completely built-up (CBU) motorcycles mostly made in China, made possible by the deregulation and liberalisation measures issued by the Indonesian government from 1997. Indonesia is now the most liberal of the three big developing countries (China, India, and Indonesia) as far as automotive liberalisation measures are concerned. Foreign ownership limitation, local content requirement, local content incentive and negative listing of the automotive industry have been lifted. Tariff barriers have been lowered, and no non-tariff barriers are in place. Even the requirement for sole agency registration in the Indonesian territory has been technically lifted, legalising the parallel import of vehicles with the same brand and specification without the obligation to support the after sales service, and in many cases by 'piggybacking' the existing after sales network of the original sole agent. These deregulation measures in the automotive industry and trade initiated by the government have brought mixed blessings to the automotive environment.

Cluster strategy in Indonesia: from SMEs to industry

SME clusters in Indonesia: first cluster strategy implementation

SMEs are facing an era of world trade liberalisation and economic globalisation. Demands are made on SMEs to improve their efficiency and productivity to adapt to and be flexible as regards markets, technology, product, management, and organisation. However, SMEs, particularly in developing countries such as Indonesia, are having difficulties in capturing these opportunities that require products with better quality, prices and good service, after sales service, larger production quantities, and homogeneous production standards (Hill 2001; Tambunan 2005). Therefore, many enterprises experience difficulties in achieving economies of scale. They also come across significant obstacles to internalising functions.

Accordingly, for Indonesian SMEs, clustering is believed to offer opportunities to engage in a wide array of domestic linkages between users and producers, and between the knowledge producing sector (universities and R&D institutes) and the goods and services producing sectors of an economy that stimulate learning and innovation (Feldman 2005). Furthermore, clustering allows Indonesian SMEs to grow in stages by sharing the costs and risks through collaboration.

Cooperating SMEs in a cluster may take advantage of external economies: presence of suppliers of raw materials, component, machinery and parts; presence of workers with sector specific skills; and presence of workshops which make or service the machinery and production tools (Tambunan 2005). Also, with the clustering of enterprises, it becomes easier for the Indonesian government, large enterprises (LEs), universities, and other development supporting agencies to provide services: such services and facilities would be very costly for the providers if given to individual enterprises in dispersed locations (Kuncoro 1996). Particularly in a country such as Indonesia where agglomerations of SMEs are located in rural areas, it is more beneficial for SMEs to form into clusters.

SMES are a clear and consistently stated Indonesian government priority. They feature prominently in key government documents, such as five-year plans (Repelita), the Broad Outlines of Government Policy (GBHN), and many official statements (Hill 2001).

In addition, the government has planned VISI IPTEK 2025 (Indonesian Knowledge and Technology Vision 2025) as the ideal outcome for the national development and prosperity of the Indonesian people (MOSR 2002). Therefore, research institutions under the state Ministry of Research and Technology and universities across Indonesia will work together. They will look at the applied and appropriate technologies for SMEs, as well as learning activities among these.

Indonesia values SMEs as significant economic drivers in addition to those drivers provided by LE activities. In 2003, SMEs accounted for 99.25 per cent of the total industry in Indonesia. Moreover, in terms of employment opportunities, SMEs provide 59.82 per cent of the country's total industrial employment. It is therefore unsurprising that SMEs receive a lot of attention in Indonesia (Tambunan 2005).

Furthermore, there is a particular current interest in SMEs in Indonesia since these firms appear to have weathered the economic crisis of 1997–1998 better than larger industrial units. This proposition appears to be true both for intra-country comparisons (large and small firms within a given country) and across economies (the Korea–Taiwan comparison) (Tambunan 2005).

The clusters were established naturally as traditional activities of local communities whose production of specific products have long been proceeding and where the workers have special skills in making such products (Tambunan 2005). Batik clusters (the traditional Indonesian textile) within the District of Java Island (Yogyakarta, Pekalongan, Surakarta and Tasikmalaya) are a long-standing example.

In 2003, 69.05 per cent of SMEs were concentrated in the Java region, compared with less than 12 per cent in the Sumatra region and less than 1 per cent for Maluku and Papua in the East region (BPS 2004). The concentration of SMEs in Borneo is less than 5 per cent. In light of these facts, the Indonesian government must admit that there is an imbalance in the distribution of economic development across regions.

Hill (1997) describes the importance of clustering not only for the development of SMEs in the cluster, but also for the development of villages/towns in Indonesia (see also Sandee and ter Wingel 2002). He gave the example of how the clustering of rattan furniture producers has absorbed an entire village in Tegal Wangi, West Java and created several small-scale industrial activities in neighbouring hamlets. Another scholar, Sandee (1994) provided similar evidence from wood furniture in Jepara, Central Java, when the growth of this cluster in the 1980s had transformed the town into a thriving commercial centre with many furniture showrooms and factories, modern hotels, new commercial banks, supermarkets and European restaurants. Therefore, clustering is indeed important for the development of SMEs as well as for a region's social and economic development.

Nevertheless, it has been found that not all the SMEs clusters within the District of Java Island are successful. Some of them have found it difficult to thrive because of market competition, particularly those enterprises which have been established in rural areas. Therefore, it is urgent for both government and universities to reach out to these rural clusters in order to get involved in building up a cluster approach (Ardani 1992; Kuncoro 1996).

Clusters for SMEs in Indonesia can be classified into four types, according to their level of development. Data from the Central Bureau of Statistics shows that the first type is *artisinal*, indicating that the process of clustering is still at an 'infant' stage.

The second type is *active*; indicating that it has developed rapidly in terms of skill improvement, technological upgrading, and successful penetration of domestic and export markets. Examples of this cluster are roof-tile clusters, metal-casting clusters, shuttle-cock clusters and shoe clusters.

The third type is *dynamic*, indicating the decisive role of leading/pioneering firms, usually larger and faster growing firms, to manage a large and differentiated set of relationships between firms and institutions within and outside the clusters. Examples of this type are clove cigarette clusters in Kudus, tea-processing clusters in Slawi, and tourism clusters in Bali. In the case of clove cigarette clusters, their products are able to outperform products from Phillip Morris and British American Tobacco (BAT). Tea-processing clusters led by big companies such as Sosro have grown to become market leaders in the Indonesian soft drink market, leaving giant Coca-Cola behind.

Clusters of the fourth type are more advanced, more developed and more complex than the previous types. There are two of these well-known cluster agglomerations in Indonesia. The first is in the Yogyakarta–Solo area with its tourism, furniture and interior decoration, metal processing, textile and leather goods, which all mutually benefit each other. The second one is Bali, known as a tourist destination, with SMEs that produce traditional handicrafts, furniture and interior goods, silver jewellery, and paintings.

Industrial estates for agglomeration: cluster strategy for industry

An industrial estate is defined as an area managed and marketed by a private or public company to offer infrastructure (electricity, water, sewage system) and in addition it may also provide a host of supporting services (permits, security) and facilities (HKI 2006).

Industrial estate companies are legal entities established under the law of the Republic of Indonesia and domiciled in Indonesia. The companies can have the following legal forms:

1 National private company (PMDN)
2 Foreign investment joint venture company (PMA)
3 State-owned enterprises (BUMN)
4 Non-facility companies (i.e. regular domestic firms)

Any company (either domestic or joint venture) can start developing an industrial estate with a minimum of ten hectares (ha), provided it has the necessary legal permits, particularly the location and acquisition permit from the local government and the industrial estate developments permit from the Ministry of Industry and Trade (HKI 2006).

Industrial zones are areas designated by the central or regional government where industrial development should be promoted and nurtured based on sector or regional priorities and plans; outside these designated zones, no other industrial sites should be permitted, in order to protect fertile agricultural land and the environment. However, many discrepancies still exist between planned and actual land use. Generally, industrial estates are only permitted in areas designated as industrial zones, which cover an approximate area of 500,000 ha in Indonesia. For example, in West Java alone, there is at least 1,000,000 ha of land that potentially could be used for industry.

In response to government regulations, industrial estates are to serve as an instrument for the allocation of industries in line with the regional spatial plan, in order to generate industrial development growth. Development of the estates should reduce the acreage of farmland and should not be established on land that is intended for the conservation of natural resources and cultural heritage.

According to the last record by the Industrial Estate Co-ordination Board (HKI 2006) there are 225 industrial estate companies across Indonesia, covering 8,051,478 ha. These industrial estates can be classified by the following criteria:

1 State-owned or public enterprise (as opposed to private sector companies)
2 Industrial estates that include Bonded Zones: export processing zones (EPZs), Private enter-port for Export Destination (EPTE)
3 Legal status of the industrial estate company: joint venture/PMA, private/ PMDN
4 Type of industrial estate: techno/science parks, specialised industrial estates

In Indonesia, industrial estates generally offer:

1 Infrastructure: roads, water supply, drainage systems, waste water systems, electricity, telecommunications, a location-based industrial environment.
2 Specific information related to the plan/site of the industrial estate.
3 Special facilities: employee housing, office space, conference rooms, fibre optic telecommunication cables, special transport services.

There are many sector industries in Indonesia such as palm oil, crumb rubber, plantation products, shipping, textiles, pharmacies, cosmetics, automotives, pulp and paper, and other heavy industries. Subsequently, Java maintains the largest regional distribution of industrial estates throughout Indonesia containing 75 per cent of all industrial estates; the province of West Java alone contains 50 per cent.

With standard infrastructure, electricity, communication, etc. becoming available throughout Java, transportation is the prime factor for locating the industry in this region. As a result, most of the industrial estates in West Java are owned by private companies, several of which cooperate with foreign companies (mainly Japanese and Korean). Five industrial estates are state-owned with three of them located in DKI Jakarta (Pulo Gadung, the EPZ Kawasan Berikat Nusantara, and Pluit Distribution Centre); the other two are Krakatau Industrial Estate Cilegon and Kujang Cikampek Industrial Estate.

The capital city, Jakarta, has four industrial estates with a total area of 1,277 ha. Due to the limited supply of industrial estate land and high prices, many investors expanded their industries to regions outside Jakarta such as Depok, Bekasi, Tangerang, Karawang, Cilegon and Serang (HKI 2006).

Subsequently, East Java is another growth region, with Greater Surabaya becoming Indonesia's second largest industrial centre. With an expanded international airport and seaport, East Java is the gateway to the source-rich regions in eastern Indonesia.

In Central Java, the city of Semarang located between Jakarta and Surabaya has also become an industrial centre. Additionally, the other Indonesian island with substantial government support is Batam, with its strategic location just across from Singapore and Malaysia. Batam has been declared a bonded area for exports in order to benefit from its strategic location.

The early years of industrial estate development were dominated by state-owned companies. The situation has now been reversed with the private sector accounting for about 80 per cent of industrial estate companies. According to HKI (2006), privately-owned industrial estates have developed better compared with state-owned industrial estates. Thus, the majority of investors (primarily foreign investors) are in favour of being located in privately-owned industrial estates since some of them are joint ventures. This is the case for most of the automotive investors who prefer to relocate their plants in Japanese-Indonesian joint ventures industrial estates (HKI 2006).

The current progress of industrial estates and its importance for cluster strategy

By its nature, culture and tradition, Indonesia is essentially an agricultural country facilitated by many natural resources. It is a tropical country of 202,087 square kilometres of land over 13,667 islands along the equatorial line.

In a government effort to industrialise the country, agro-industries were used to start the industrialisation process. The country, however, is open to other kinds of industry based in Indonesia that can utilise the potential work force (and market) coming from the 238 million population and the other available natural (mining) resources.

The development of industries happened during very high economic growth (about 6.0–7.5 per cent annually), which has boosted the country's export volume and value especially in the past 20 years, with the exception of the crash in July 1997 until now. This rapid growth has brought change and challenges to every level of Indonesian society. But these have brought a high cost to the environment.

It is widely recognised that the impressive increase in private investment had something to do with the government's efforts to create a conducive investment climate for both domestic and foreign investors. Following the issue of Law No. 1 of 1967 on Foreign Investment (PMA) (amended by Law No. 11 of 1970), and Law No. 6 of 1968 on Domestic Investment (PMDN) (amended by Law No. 12 of 1970), several policies were issued to improve the investment climate in Indonesia.

In spite of all the concerns and problems faced by Indonesia, the steady increase in the value of PMA investment and the development of infrastructure by the private sector should increase the opportunities to tap the market for industrial estates. Government policy regulating industrial property development and foreign investment has also fostered the growth of industrial real estate activities.

Industrial estates were introduced in Indonesia in the early 1970s. As in other countries, the basic idea is to facilitate the establishment and operation of factories by providing ready-to-use industrial complexes.

Presently, in Java, there are 61 industrial estates with a land area of around 18,000 ha. Most are privately owned although there are a number of state-owned estates (HKI 2006).

Indonesian industrial estate infrastructures include security management and community development, flood free areas, industrial roads for 8 tonne axle loads and 8 m width per lane, access roads for cargo along with access for passengers to main ports and residential areas, reliable electricity supply, established telecommunications service and waste treatment plants. Meanwhile, industrial estate facilities provide access for EPZs, bonded warehouses, transportation and amenities. Subsequently, industrial estate services include all the legal services for investment administration, export–import services, and maintenance services including security management for the factory.

Industrial estates are properly designed for manufacturing activities; potential sites were studied carefully by experts in many aspects, e.g. rainfall level, environmental impact, transportation volume generated, social and cultural impacts, and whether legally approved as a manufacturing area. Industrial estate management must ensure that every tenant follows industrial estate regulations.

The development of industrial estates particularly in the Java region (i.e. Jabodetabek) has been growing, which is considered as the most attractive area for investment in Indonesia. There are several industrial estate developments in operation by both domestic and foreign investors which have already been selected as the location for thousands of foreign and domestic manufacturing industries. Manufacturing and industry operating in this region ranges from electronics and telecommunications to automotive components and parts industries, food and beverages, ceramic and glass industries, chemical industries, and apparel (Jababeka 2005; HKI 2006).

Through its deregulation policy, the government has eased restrictions on FDI, opening Indonesia up to multinational industrial companies, the primary buyers of industrial land. In addition, in 1995 the government began restricting industrial development in areas west of Jakarta, encouraging development in the Jabodetabek region where the majority of the industrial estates are located.

The advantages of being located in industrial estates for manufacturers have been felt by both foreign and local investors, as the industrial estate is properly designed for manufacturing activities. In contrast, investors who are not located in an industrial estate have faced disruption to their productivity and efficiency related to their manufacturing activities (HKI 2006).

The partnership between SMEs and MNEs in industrial estates has become an ongoing agenda since Presidential Decree no. 41/1996. This synergy has been nurtured in particular for the automotive industry as a result of increased Japanese automotive investment (Kuncoro 2002).

Subsequently, the cluster strategy initiated by the Indonesian government has attacked the Japanese automotive industry as most of the private industrial estates are also part of Indonesian–Japanese joint ventures (HKI 2006).

Therefore, it is expected that with an improvement in stability in politics, as well as macroeconomic fundamentals, Indonesia would be able to sustain its capacity to attract investors and ultimately to continue its economic recovery.

The organisation of supply systems plays an important role in the competitive advantage of Japanese automotive manufacturers (Narula and Dunning 2002). This emphasises the close relationship between the automaker and its supplier both in Japan and overseas. On the other hand, it has been observed that Japanese automakers share more information with their suppliers and that the coordination of tasks is strong, and they are therefore willing to make relation-specific investments (Fujimoto 2007). Thus, Japanese supply networks pointed out that long-term, mutually beneficial relations influenced the choice of location for their investment.

The Indonesian government initiated the setting up of industrial estates, later labelled 'industrial clusters', to make the existence of stakeholders easier, for both domestic and international investors, by providing all necessary infrastructure, facilities and housing in one safe location, and at a reasonable cost, thus providing a secure base for industry and manufacturing.

Another rationale behind the creation of industrial estates is to emulate the success of SME clusters across the Indonesian region. If most SME clusters naturally occur due to geographical proximity, along with the skill-level and mobility of the labour, there are two causes for the formation of industrial clusters (Ardani 1992). First is the organic cluster whereby the firms are organically agglomerated into certain geographical areas followed by the trinity of externalities. Second is the artificial cluster, whereby the firms are designated a location in certain industrial estates.

In an attempt to advance the economic condition through industrial policy, the Indonesian government divided *Indonesia Bangun Industri 2025* (Indonesian Industrial Growth 2025) into a two term setting-up plan; middle term (2004–2009) and long term (2010–2025). The middle-term industrial objectives were set as follows:

1 Improving the level of employment in the industrial sector
2 Improving the Indonesian export and empowerment of the national market
3 Providing significant contribution to the national economy
4 Supporting the development of infrastructure
5 Improving technological ability
6 Improving the core of industrial structure and product diversification
7 Improving the balance of industrial distribution across the nation

In addition, the long-term industrial objectives are:

1 Strengthening the core manufacturing industry in order to be a world class industry
2 Improving the role of priority industry as prime mover in national economy
3 Improving the SMEs' role as indigenous enterprises to play an active part in the national economy along with large manufacturing industry

In order to achieve the objectives above, the government has initiated an Operational Strategy:

1 **Developing a supportive business environment**
   More cooperation with the relevant institution is needed to develop infrastructure in the areas where existing or potential industry takes place, improving human resource management specifically in production engineering as well as business management, developing industrial zones whereby there are some industrial districts developing R&D services, calibration services, quality management based on ISO 9000 and ISO 14000 standards, and other relevant manufacturing services to qualify for the highest quality in production; enhancing the relevant industrial policy in order to support the existence of manufacturing industry; better synchronisation between industrial policy, investment, and trading policy to support opportunities conducive to business in Indonesia; and paring down the bureaucratic process for investing in Indonesia.

2 **Setting up industrial development through the implementation of a cluster strategy for primary industry**
   The Indonesian government has set up five priorities for industrial clusters as prime movers in developing industry. The framework used to set up priority industry is based on international competitive advantage and the Indonesian potential in developing industrial conditions.

   The working definition of clusters in this research is *a geographical concentration of related industries and institutions*. In Indonesian, *sentra* is a similar concept, as it is defined as a geographical concentration of manufacturers in the same sector. The five priority clusters are:

   1 Agro industry
   2 Transportation (automotive, ship, aviation, train) and machinery industry
   3 Electronic and information-computing industry
   4 Basic manufacturing industry
   5 Selected SMEs to be involved with large manufacturing industry along with other supporting industrial cluster and related industries, mainly in the agriculture, food processing, shoes, sea and marine, textile, rattan and timber, pulp, palm oil, electric equipment, and chemical industries.

3 **Setting up eco-regional priorities**
   This aims to balance industrial distribution across the nation so it is not concentrated in the Java region. The other reason is external proximity whereby raw material, labour, and variable cost will be cheaper for industry use.

4 **Improvement of innovation ability, particularly in industrial and production engineering, and business management**
   Specifically, the enhancing of R&D in industry involved with design, production engineering, plan construction, and equipment fabrication.

At an early stage, most automotive makers grew organically in the north of Jakarta. They clustered themselves in this area as there were no designated and well-managed industrial estates from the government (HKI 2006). In conjunction with the typical Japanese *keiretsu* network, most of the automakers in north Jakarta were Japanese, building up their value chain network. As a result, most of the suppliers were located in the same areas (Kuncoro 1996, 2002).

Nevertheless, when the area grew crowded, along with new regulations from the DKI Jakarta Urban Planning Division, the solution was to take up opportunities provided by new regulations from the Indonesian Investment Coordinating Board, using industrial estates to accelerate the growth of foreign and local investment.

Today there are a total of 40 industrial clusters currently operating in Indonesia. Of these, 32 are located on the island of Java, the country's most densely populated region. Of the remainder, three industrial clusters are located on Batam Island, one on Bintan Island, three on Sumatra and one in Makassar, South Sulawesi (Jababeka 2005).

Indonesia's industrial clusters provide foreign investors with the infrastructure and facilities necessary for setting up industrial and manufacturing operations. In addition, they afford a high degree of security and an increasing number also have the advantage of integrated housing, thus providing accommodation for both expatriates and work force. As Indonesia's economy continues to improve, the number of industrial clusters will also increase.

Seventeen of Indonesia's industrial clusters are sited in the Greater Jakarta metropolitan area referred to as 'Jabodetabek', being an acronym of Jakarta, and its neighbouring cities of Bogor, Depok, Bekasi, Tangerang and Karawang. Many investors prefer to be close to the capital city because of its proximity to the Soekarno-Hatta International Airport and Tanjung Priok, Indonesia's largest harbour complex (HKI 2006).

In conjunction with the growth of industry, manufacturing activities need specific infrastructures, services, and facilities to support their typical manufacturing production activities. Therefore, industrial estates are needed to provide such support, specifically for manufacturers who require a complete infrastructure and facilities, a one-stop service for any necessary permits for manufacturing activities, a supportive attitude from surrounding communities, and the support of both central and local government and other related institutions for productivity improvement.

In view of that, the development of the automotive and component and part industries is of growing importance to this area, with the establishment of new investment in several manufacturing facilities such as in Cibitung (West Java) for the manufacturing of Honda and Suzuki motorcycles with an annual production capacity of 1,000,000 units respectively. In Cikarang, there will be new investment from Malaysia for the production of Proton cars (HKI 2006).

Additionally, in West Karawang, there are Toyota and Daihatsu car manufacturers that have been operating to full capacity and new investment by Yamaha motorcycles to produce a motorcycle with an annual capacity of 1,000,000 units.

In East Karawang, the manufacturing facilities for the Honda CRV also operate to their full capacity. In Cikampek, there are located manufacturers of machines for Honda cars and Nissan trucks.

It is believed that there would be significant growth of foreign investors coming into this corridor area as vendors and suppliers to the automotive manufacturing industry. However, it would require government support to make a credible commitment to undertake the necessary programmes. Many social and infrastructure programmes that were planned more than 15 years ago still remained unrealised.

In order to restore investor confidence, the government must vigorously address the above issues. Any efforts by the government to improve the current infrastructure, policy, and tax-fiscal conditions will be perceived by Japanese investors as a signal of the government's willingness to work with foreign investors. Thus, it can be a good signal to foreign investors who may be considering Indonesia as a location for their investment.

Conclusion

The development of the country's automotive industry is stimulated by government policies regulating the sector, as well as technological advances and prevailing economic conditions. Indonesia's automotive industry is a highly export-intensive industry (Okada 1983; Sugiyama and Fujimoto 2000; Tarmidi 2004). Over time, the industry has gradually reduced import intensity, as more and more components and engines are produced domestically. Another feature is the domination of Japanese automotive MNEs as a source of knowledge and technology transfer. This is consistent with Japan's dominance in Indonesia's automotive market.

The mood of the automotive clusters in Indonesia has been characterised by the trend towards liberalisation (Narula 2003; Rasiah 2004; Coe et al. 2007). Under these circumstances, the improvement of competitiveness becomes the key factor for survival. Since the automotive markets in developed countries reached maturity in the 1990s, the focus of the global automotive market has shifted to emerging markets such as Asia and Latin America (Dowling and Cheang 2000; Rasiah 2005; Rasiah et al. 2008). In conjunction with that, the need to be more environmentally friendly, and for high standards of safety, requires investment in R&D, another challenge for the automotive makers in Indonesian automotive clusters.

Nevertheless, to be able to strengthen these alliances, the automotive clusters in Indonesia and other ASEAN countries should possess some strengths in general and specific aspects, such as brand loyalty, competitiveness in price, production bases in specific locations, marketing networks, technological capabilities, and respectable market share. Therefore, it is important to emphasise the potential of the automotive industry as a contributor to future growth of the ASEAN market.

Specific to the Indonesian condition is the aftermath of the financial crisis. Improved exports and strong domestic demand will aid the recovery from the

Global and national environments 81

financial crisis (MOTIRI 2005). Automotive exports comprise finished products and parts exports. Finished automotive exports from Indonesia are limited to commercial vehicles (i.e. Kijang) which are exported to Southeast Asian countries, Latin America and Africa (Tarmidi 2004; TMI 2007).

As regards cluster strategy, clusters are regarded by the Indonesian government as a suitable and sensible approach, with SMEs and MNEs located in diverse areas. Therefore, it will be effective and efficient for universities and other institutions if they engage and develop these clusters by providing technical assistance. It is agreed among observers that Indonesia will have a much better future once the current financial crisis is over. Therefore, industrial estates have to prepare themselves to build the eco-friendly infrastructures required to tap the upcoming opportunity, despite the fact that the challenges ahead are far more complex, due to the close connection between the global economic situation and the manufacturing sector.

In this chapter, I have outlined the effect of macroeconomics on national industry and trade policy in Indonesia, particularly in the automotive sector, and on the importance of clusters as a location for MNEs. Moreover, Indonesian automotive clusters are related to ASEAN automotive clusters, as they mostly consist of Japanese MNE transplants. Therefore, forging networks among them is a crucial agenda for the policymaker in the ASEAN region.

The next chapter elaborates the research questions and methodologies, outlining how the case studies were conducted: intensive fieldwork in Indonesia, the process of data collection, data analysis, and reflection on the case studies in Indonesian automotive clusters.

5 Methodologies

This chapter details the methodologies in relation to the questions posed in Chapter 1:

1   **Context:** How has the global-local network in the automotive industry impacted on the automotive cluster in Java?
2   **Process:** How have the key actors (Toyota and Honda) been involved in fostering technological change and knowledge-technology transfer in the Indonesian automotive cluster (through localisation project and suppliers)?
3   **Outcomes:** Using two case studies of Toyota and Honda as multinational enterprises (MNEs) in the automotive industry, how have they contributed to the development of the Indonesian automotive cluster (new knowledge, new skills)?
4   **Recommendation:** How can the cluster policy be improved and what lessons can be learned?

In order to answer the research questions, I have used an *explanatory* case study method, complemented by two other types, *exploratory* and *descriptive* case studies, with the purpose of analysing and reporting the wider picture of Indonesian automotive cluster studies.

Any researcher must apply great care in designing and carrying out a case study to overcome traditional criticisms of the method (Stake 2000; Osland and Osland 2001; Yin 2003a, 2003b; Silverman 2004). I decided to use the explanatory case study method after taking into account four key issues, namely research questions, my skills and attitude, costs or budget available, and time available (taking into consideration the target date for PhD completion).

In view of the above, in order to complete the data collection, fieldwork in Indonesia was carried out, and the findings and discussion are elaborated in Chapters 6, 7, 8 and 9. I visited the people, sites, and organisations to interview, observe and record behaviour in its natural setting. Therefore, the qualitative process used in this research is *inductive* whereby the researcher builds abstractions, concepts, findings, and conclusions from details. Further details of the research design underpinning the fieldwork is discussed, to complement the theoretical framework of Chapter 2 and Chapter 3 along with the extended explanation in Chapter 4 on the macroeconomic background and industrial policies, particularly the automotive policy and cluster strategy in Indonesia. Furthermore, the case studies have been conducted on the basis of contextual data on cluster policies, particularly in a developing country, as well as Japanese business networks in the automotive industry along with lean production as the core of the Japanese automotive manufacturing system.

Explanatory case study for the Indonesian automotive cluster

The use of the term 'case study' draws attention to the question of what can be learned from a case (Stake 2000; Yin 2003a). Therefore, a case study is defined as a high-level approach to the research that determines much of the detailed work, specifically fieldwork and ongoing observation and correspondence (Denscombe 2000; Mason 2002; Silverman 2005).

This research uses the explanatory case study method to answer the proposed research questions in Chapter 1. Explanatory case study is chosen as it allows the researcher to concentrate on a specific instance, that is, the Indonesian automotive cluster, in an attempt to identify detailed interactive processes. An explanatory case study is a sensible method to use in international business and management studies (Buckley and Chapman 1996; Buckley 2002; Marschan-Piekkari and Welch 2004), where cross cultural understandings (such as Japanese business characteristics in the automotive *keiretsu*) are a necessary element to be taken into account. However, the use of the explanatory case study method is arguably the most difficult and the most frequently challenged (Stake 2000; Yin 2003a, 2003b).

Additionally, an explanatory case study is the preferred strategy when 'how' or 'why' frames the research questions and when the researcher has little control over events because the focus is on a contemporary phenomenon within some real-life context (Maxwell 1996; Yin 2003a, 2003b; Seale *et al.* 2004; Silverman 2005). In view of this, the case study method underpins this research, which is concerned primarily with *process*, *meaning*, and *understanding* gained through overseas fieldwork in Indonesia. That is the rationale behind the methodology for this research.

In this research, each case study seeks to explain 'how' and 'why' the cluster strategy affected the development of cars and motorcycles in the Indonesian automotive cluster, along with the question of 'how' the knowledge-technology transfer has progressed in each of the industries. Therefore, embedded in the explanation is a potential historical and technological path dependency, whereby the case studies of Toyota Motor Manufacturing Indonesia (TMMIN) and Astra Honda Motor (AHM), demonstrate that the companies seem to be making inroads into current attribution problems.

The case study is both the process of learning about the case and the product of its learning. Therefore, after considering the three conditions, namely the type of research question, the control the researcher has over actual behavioural events and the focus on contemporary as opposed to historical phenomena, the

84 Methodologies

case study method is chosen because it is expected to advance the understanding of the research phenomenon.

Subsequently, I chose empirical research which draws on fieldwork experience and participant observation where I had been allowed to observe the primary evidence (the production and development meeting of the engineers and the learning process among the engineers) in order to understand the phenomenon under study. Additionally, I interviewed relevant employees' representatives from the following affiliations in order to collect new evidence, which tended to be qualitative:

1   TMMIN and its suppliers (one auto maker and five auto parts suppliers)
2   AHM Indonesia and its suppliers (one auto maker and six auto parts suppliers)
3   Industrial Estate Developers where the automotive industry is located (three developers)
4   University with links to the automotive industry
5   Government bodies

Accordingly, I approached the work as an interpretivist. The interpretivist or qualitative approach has been chosen with the following considerations in mind:

1   The aim is to provide an in-depth understanding of the Indonesian automotive cluster and the processes within it.
2   Two case studies in car and motorcycle manufacture are deliberately selected against particular criteria.
3   Evidence collection methods typically involve close contact between the researcher and the subjects being studied.
4   Evidence is detailed, information rich and extensive.
5   Analysis is generally open to emergent concepts and ideas.

Therefore, I was aware of these key issues and their potentially significant influences on the research. In practice, I worked hard to overcome the problems of doing case study research: case study research is remarkably hard, even though case studies have traditionally been considered to be 'soft' research.

Research setting and research design continued with data collection

My experience over the PhD phase of conducting case study research certainly confirms Yin's view that case study research is significantly hard (Yin 2003a, 2003b). It is not just a matter of going to do field work to visit the companies, but also interviewing respondents, and carrying out observations in the field site followed by writing up the results (Wolcott 1995; Baxter and Eyles 1997; Ryen 2002).

In reality, case study research requires clear research questions, a thorough understanding of the existing literature, a well-formulated research design with sound theoretical underpinning, and, above all, excellent language skills.

Language skills relate to the fact that the researcher must be able to synthesize large amounts of diverse data, such as interview notes and transcripts, documents, observations of meetings, and documentary evidence; then from all these data, produce theoretically informed and convincingly argued conclusions (Maxwell 1996; Stake 2000; Silverman 2004). More importantly, the researcher must be able to communicate with both the subjects of the research and the readers of the resulting document.

Originally, in September 2005, I intended to focus only on export-oriented manufacturing clusters in Indonesia, due to the fact that most of the industrial estates across the Java region have turned to export-oriented industry based on several prominent sectors, namely electronic clusters, automotive clusters, food-processing clusters, apparel clusters, and SME clusters.

Nevertheless, as I developed my understanding of the cluster theoretical framework and its application for regional studies along with the international business and management literature review on knowledge transfer of the Japanese automotive MNEs, I deliberately decided to focus more on sectoral industry. This gave rise to two options, either the electronic cluster or the automotive cluster. In this respect, I carried out preliminary analysis on both options.

For the electronic cluster, I reviewed the detailed literature on electronic cluster and network studies in Southeast Asia, relevant journals, sufficient textbooks, as well as personal experience gained when working for PANASONIC in Batam Island, the favourite electronic cluster for Indonesia, Singapore, and Malaysia (the 'golden triangle' electronic region located on Sumatra Island). Nevertheless, the data access to those electronic MNEs in Batam Island was limited along with poorly supported research resources from the relevant electronic industry associations.

Additionally, due to current volatile global economic conditions after the Asian financial crisis, the electronic industry gives an impression of bleak prospects for long-term foreign direct investment (FDI; Ernst 1994; Ginzburg and Simonazzi 2004; Rasiah 2005), where MNE investors mainly transferred limited production capability and knowledge to the host country.

Due to the nature of its manufacturing activity, the electronic industry is a labour-intensive industry in contrast to the automotive industry (Gammeltoft and Aminullah 2004). Subsequently, the latest government industrial policy states that electronic industry is not the main prime mover for the Indonesian Vision 2025 of a more advanced industrial nation (MOSR 2002; MOTIRI 2005). However, the Indonesian government will still pay attention to this industry as it generates labour employability at a significant scale (Aswicahyono and Feridhanusetyawan 2004; Rasiah 2005).

On the other hand, conducting research on the Indonesian automotive industry from a cluster perspective was more promising despite the lack of literature in review. In conjunction with the cluster strategy, the pattern of MNE location (specifically for the Japanese automotive MNEs) has spread across the Java region (mainly in West Java). Thus, the chance to explore more about the cluster process and its interaction is visible. Furthermore, the automotive industry has existed in Indonesia since 1927; this industry has experienced longer trajectories and had time to encounter various problems with Indonesian industrial policy, which is an important aspect for studying the life cycle of industry in Indonesia (MOSR 2002; MOTIRI 2005).

Subsequently, in relation to the research question on the process of knowledge-technological transfer and the actors involved in Chapter 1, the Indonesian automotive industry provided promising research potential. The automotive industry is located in clusters and it has become more established and developed compared to other type of industry clusters, along with the learning process of the engineers within the automotive cluster (known as *jishuken* activity) and the suppliers in another. This postulation was the result of the researcher's preliminary fieldwork preparation in 2005–2006 and 2006–2007, together with ongoing correspondence with the relevant respondents in Indonesia.

The fieldwork preparation (including pilot study) took six months in 2006. I carried out a pilot study to test out three different semi-structured interview questions for industrial estates, government bodies, the automotive players, and any other relevant respondents. The pilot study was carried out in two ways: internal discussion and external test.

Internal discussion means that the questions were discussed with some other senior scholars in the Business School and my supervisors. External tests were done by sending out the interview questions to the interviewees, the list of which was collated as a result of ongoing email correspondence with various people from the Indonesian academic/research institutions, government bodies, the automotive players in the Indonesian automotive cluster, and senior scholars met at UK and international conferences, seminars and research courses.

Consequently, the interview questions were revised as a result of the pilot study. Additional interviewees were recruited during the fieldwork in Indonesia. In total, fieldwork was undertaken with 34 interviewees.

Following fieldwork preparation, I compiled the relevant tools to conduct fieldwork in Indonesia. Since the data access was so critical and a significant step for fieldwork, I sent formal invitations and research proposals by email in order carry out field work in several chosen industrial estates of the automotive industry. Along with email correspondence, I conducted several phone interviews using Skype, to ameliorate the effects of time difference and distance between the UK and Indonesia. The conversations were also recorded. This process took three months to produce a fixed date for field work in Indonesia.

The next step was to approach government bodies in Indonesia. Due to the nature of the research questions in Chapter 1, I approached respondents from the Indonesian Investment Coordinating Board (BKPM), the Ministry of Science and Technology, the Ministry of Trade and Industry, relevant universities with project knowledge on technology transfer with the automotive industry, the Indonesian Industrial Estate Association (HKI) and relevant business associations.

These governmental bodies are essential in explaining the rationale behind the cluster strategy for industry and its development. Due to difficulties in gaining access to these bodies, I sent several preliminary research proposals via email, with follow-up phone conversations to convince them personally. As the process to get fixed dates to visit them was quite slow due to bureaucracy, in the end, after four months of dealing with emails and phone conversations, I fixed dates for my research visit to Indonesia. However, not all the governmental bodies could agree to visits within the geographical and time constraints. Therefore, I conducted ongoing email correspondence and compiled documentary evidence.

I also approached relevant academics in the university who had been involved with automotive industry studies and who were familiar with Indonesian science and technology policy and industrial policy. As a result, I came up with a snowball process of interview and ongoing email correspondence to reach the appropriate academic circles.

Building up research networks among the Indonesian established academy was another challenge for me, as I have been a self-funded researcher based in the UK since 2004. Therefore, finding the right theme and time to match their research interest was essential. This process has been ongoing since September 2005. With the help of various scholars from Indonesian Agency for the Assessment and Application of Technology (BPPT), Indonesian Institutes of Sciences (LIPI), Indonesian Centre for Strategic and International Studies (CSIS), University of Indonesia, Institute of Technology Bandung (ITB), I eventually succeeded in building a 'virtual research peer-group' to develop the research design prior to the fieldwork. In the end, it was a time consuming process but one worth doing.

The last important actor in this research is the automotive industry itself. I carried out preliminary studies on the Indonesian automotive industry, specifically to find out who has the leading market share and who the significant players are in the knowledge-technology transfer within Indonesia's car and motorcycle industry. The outcomes of this preliminary research can be found in Chapters 2 to 4. The Japanese automotive industry is a complex tiered chain of suppliers not easily accessed. As result, I had to narrow down the list of potential automotive players for this research. Based on the preliminary research, I chose Toyota as the significant player in the car industry and Honda as the significant player in the motorcycle industry.

As with the other actors, I conducted ongoing email correspondence along with phone interviews to fix the schedule for fieldwork in Indonesia. After four months of time consuming correspondence, I set dates for conducting semi-structured interviews and observations at TMMIN and AHM, along with selected suppliers in the Java region.

Despite all the preparation and fixed dates between the researcher and the respondents, there was always the possibility of missing those dates due to bureaucracy, hectic manufacturing schedules, and other logistical matters. Hence, a contingency plan was needed; I booked time to have more than one visit to each of the respondents. In the mean time, prior to the Indonesian fieldwork in April 2007, I maintained contact with the respondents through emails and phone calls

88 *Methodologies*

in order to keep their interest in being research participants. Furthermore, I asked the respondents to suggest other relevant people who might be useful and knowledgeable in gaining relevant research findings. Having more respondents was important, considering the time, budget, and logistical issues when conducting fieldwork overseas.

I then composed a research design to underpin the fieldwork. The development of this research design is a difficult part of doing case studies. Unlike other research strategies, a comprehensive 'catalogue' of research design for case studies has yet to be developed (Yin 2003a, 2003b; Silverman 2005).

A research design is a logical plan for getting from here to there, where 'here' might be defined as the initial set of questions to be answered and 'there' might be found a number of major steps, including the data collection and data analysis (Mason 2002; Yin 2003a, 2003b; Seale *et al.* 2004). Another way of thinking about a research design is as a 'blueprint' of research, dealing with at least four concerns: what questions to study, what data are relevant (i.e. sources), what data to collect, and how to analyse the results (Marschan-Piekkari and Welch 2004). Therefore, research design was more than just a work plan for my Indonesian fieldwork.

In this sense, research design is significantly helpful for dealing with both logical problems and logistical problems in order to minimise potential flaws in overseas fieldwork. The fieldwork undertaken in Indonesia was time consuming and costly, and I believe that having an appropriate research design in the first place was of crucial significance.

As a result, the latest design for this research has resulted in six components, namely: research questions, case studies, methods, sources, analysis, and the organisation of the findings. There are three tables of research design, based on the research questions (RQ) in Chapter 1 (the last research question has been omitted).

Figure 5.1 is the research design for the first RQ, regarding outcomes.

Followed by two case studies, first is the cluster strategy for Indonesian industrial policy, second is development of cluster strategy. Each of them uses various sources, methods, and analysis.

For case study 1 on the cluster strategy for Indonesian industrial policy, I used phone interviews to contact the relevant university, HKI and industrial developers. In fact, ongoing phone and email correspondence took up to four months prior to undertaking the fieldwork in Indonesia.

During the Indonesian fieldwork, I conducted semi-structured interviews with the government (i.e. BKPM), HKI, and the university, and followed up with phone and email correspondence when the researcher I returned to the UK. The last source used was documentary evidence.

During the email correspondence and fieldwork visits, I was provided by the respondents with relevant documentation on Indonesian industrial policy, industrial estate development and the cluster policy document. The outcome is located in Chapter 4, which provides further Indonesian context. Due to the scarcity of qualified research papers on Indonesian cluster studies and Indonesian industries (including automotive), I have tried to build a framework for the

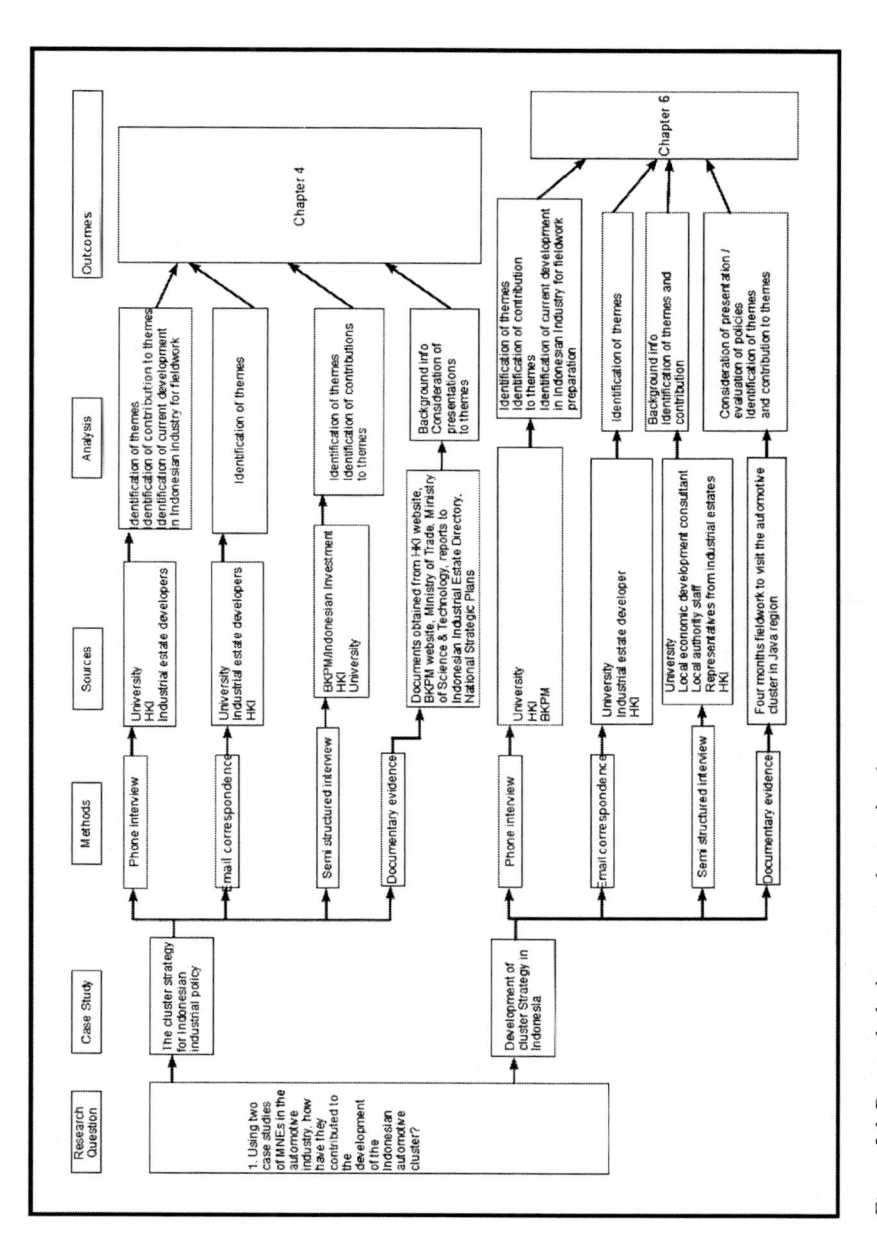

Figure 5.1 Research design: research question 1.

90 Methodologies

'Indonesian picture' prior to and after the fieldwork. Obviously, regular changes happened during the analysis process.

In relation to the development of cluster strategy in Indonesia, similar methods and sources were used. The first step was to hold phone interviews, alongside ongoing email correspondence with the university, HKI, and BKPM. Thus, during the fieldwork visit, I undertook semi-structured interviews with those respondents together with constant participant observation in the automotive cluster for four months. The outcome is located in Chapter 6.

Figure 5.2 is the research design for the second RQ, regarding process. There are two case studies. The following table elaborates the details of the case studies.

For the TMMIN case study, I used five methods to approach the company and suppliers: phone interview, interview meetings, documentary evidence, plant tours during the Indonesian fieldwork, and review by respondents. In an attempt to establish contact with TMMIN, I used my network in the engineer learning group in Toyota Group (known as *jishuken* in Japanese). It turned into a snowball process, whereby existing study subjects recruit future subjects from among their acquaintances, allowing the interview process to grow and gain momentum, like a rolling snowball. This approach helped me to secure contact with, and eventually interview, TMMIN and its suppliers.

Additionally, I attended a visit by the learning group from Toyota to its supplier group to understand comprehensively the knowledge-technology transfer process. I was able to observe both the internal process (in Toyota's plants) and the external process (in supplier's plants). Therefore, following a meticulous notetaking and recording process, and during the preparation of draft chapters and relevant conference papers, the relevant respondents from Toyota have been able to cross-check the validity and the coherence of the data collection. The outcome is located in Chapter 7.

The second case study is AHM. Similar to Toyota, to access data in Honda I used five methods: phone interview and occasional email correspondence, semi-structured interviews, documentary evidence, participant observation (of the production and development meeting of the engineers and the learning process among the engineers in *jishuken* activity), and review by respondents. Prior to the Indonesian fieldwork, I held phone interviews to set the dates for visiting and interviewing Honda's representatives. In fact, establishing contact with Honda was partly enabled by the *jishuken* network in Toyota. Some of the engineers in *jishuken* put me contact with AHM; the remainder of interviewees were gathered again by a snowball process, whereby a few potential respondents were contacted and asked whether they knew anyone suitable for the research.

During four months of Indonesian fieldwork, I conducted semi-structured interviews with Honda's representatives and its suppliers, and attended plant tours in their motorcycle and component parts plants. Again, to cross-check the validity and coherence of the draft chapters and relevant conference papers, respondents were asked to review the resulting content. The outcome from this element of the research is located in Chapter 8.

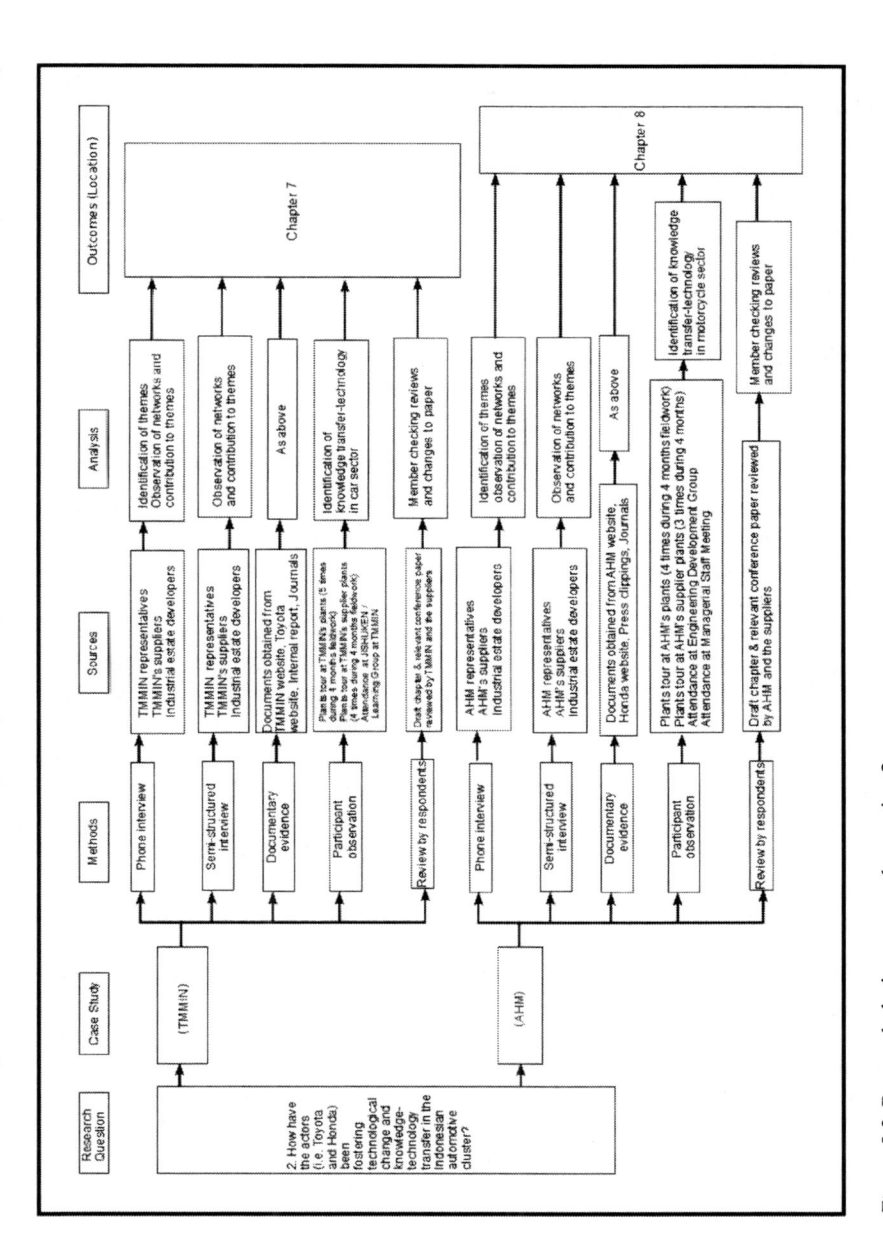

Figure 5.2 Research design: research question 2.

92 Methodologies

Figure 5.3 is the research design based on the third RQ, which looks at context. There are two case studies: Toyota and Honda. Toyota represents the car manufacture industry and Honda represents the motorcycle manufacture industry. For both, I used the same three methods, namely semi-structured interview, documentary evidence, and respondent review. The semi-structured interviews took place during four months of fieldwork in Indonesia. Documentary evidence was gathered before and after the fieldwork, against the context of a rapidly changeable automotive industry. I had to keep up to date with the latest developments in both Toyota and Honda, as well as with automotive industry research and commercial news.

Subsequently, I have remained in touch with both Toyota's representative and Honda's representative, and have appreciated their input on earlier chapter drafts and relevant conference papers. However, to avoid research bias, I also asked the relevant respondents based at the university in Indonesia to review the resulting research. The outcome is located in Chapter 9.

Fieldwork reflection on the value of explanatory case studies research

In the Indonesian fieldwork, interviewing each organisation and recording their responses would simply illustrate that each of them holds a different opinion and perspective on cluster strategies and on the development of the Indonesian automotive cluster.

When I talked to the automotive industry and to government bodies, they each gave different opinions and suggested that I should pass on various comments and requests to the other organisations (i.e. university, industrial estate developers). I did this, but in order to communicate with each organisation, it was necessary to understand their different standpoints. In this case, simply taking their responses at face value would not have enabled me to understand fully what was happening.

Nevertheless, I regard this experience as inevitable in all case studies, as communication is a two-way process, and interviewing people and asking questions can raise issues that have not previously been considered or discussed within the organisation. Eventually, by trying to get to the meaning behind the words through transcriptions of the interviews, I was able to get a deeper understanding of the case.

It is the nature of this explanatory case study research that it seeks to develop detailed and comprehensive information on the subject in question, which generally requires a combination of empirical data and qualitative opinions from key actors obtained through interview.

In the course of this research, TMMIN and AHM, real estate developers, universities, and government bodies were interviewed, generally using semi-structured interviews. Additionally, direct observations at the automotive plants across different clusters in both the components firms and the assemblers were extremely useful in gaining an understanding of the nature of the automotive manufacturing production system (lean production and the Just in Time, JIT, process).

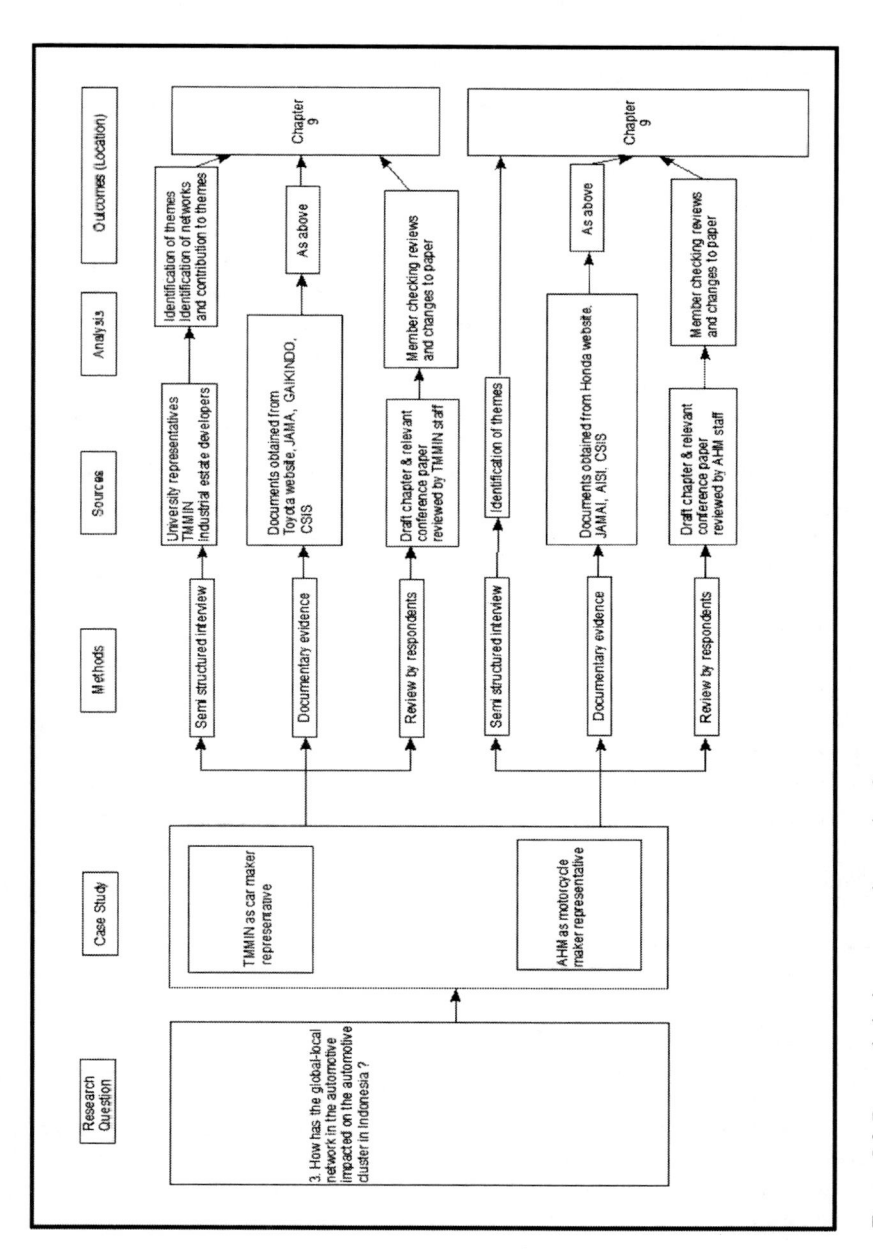

Figure 5.3 Research design: research question 3.

At later stages, detailed profiles of firms were developed by using company reports, company products, process information brochures, financial and trade directories, site visits, and interviews. In order to compare and to contrast the standpoints of the actors involved, I interviewed relevant representatives from the shop floor (i.e. production lines) prior to the managerial interviews; this was based on the assumption that on a daily basis, the shop floor might yield significant information not necessarily available at managerial level.

In other words, the managerial staff might be ignorant or mistaken about the 'reality on the ground', an observation supported by Holm *et al.* (1995) and Andersson *et al.* (2001). Thus, at the first stage, the interviews were based on the shop floor conducted primarily with engineers, line leaders, and supervisors. Later on, I interviewed managerial level representatives and directors. As a result, I could address issues such as the extent to which management and shop floor share the same understanding of work processes, strategic choices, and so on.

These same methods allow the aggregation of statistical material derived from governmental bodies and relevant business organisations in the automotive industry. I am indebted to them as a whole for the openness with which I was received; in practice only a handful of the many firms I contacted felt unable to contribute to the research. In part, this is due to the careful selection of interview targets and the assiduous application of the right procedures to obtain interviews, but equally reflects an industry willing, even desperate, to learn. Therefore, the legitimacy and neutrality of the researcher's status as a PhD researcher from a business school certainly helped, but without the cooperation of these firms and other actors involved in the interview process, the research would not have been possible.

Explanatory case study research in this thesis also involved widespread reading to gain a 'feel' for the subject of research. In this case, I felt it was important to understand the underlying production and process technology involved in the automotive industry, as well as to keep up to date with ongoing changes and news in the global automotive industry as a whole. Therefore, for instance, on explaining the knowledge transfer and knowledge flow within and outside of the automotive cluster, I felt compelled to devote attention to the production system and supply chain of the automotive in detail. This is particularly true because I am interested in the catching-up process of technology, knowledge, and organisation from FDI to the local Indonesian industry and local economy, in a framework of cluster strategy.

Ultimately, a great deal of this material, derived from specialist journals, governmental publications, consultancy reports, newspapers, academic journals, textbooks, and interviews, are not used directly. Instead, the information obtained together with that gathered through conversations with supervisors, colleagues, and other established researchers during conferences or relevant research meetings/workshops, combines to provide a multifaceted understanding of the sector.

The task of the researcher here is one of selection, in which relevant information is presented to support the arguments offered. As such, the work is inevitably biased. I have a bias towards English, Japanese, and Indonesian language publications: for instance, the researcher's implicit knowledge of the industry is strongest in the Indonesian context. Such biases cannot be eliminated, only acknowledged and allowances made for them. Therefore, I have endeavoured to keep my perspective as neutral as possible.

In view of that, this explanatory case study research process cannot be considered linear, when a priori theory informs empirical research which subsequently proves or disproves generalised statements. Rather, the two areas of theory and practice (experimental empirical reality) interact over time. Theory is constantly adjusted in the face of new information, which in turn shapes the way in which the researcher investigates and interprets the real word. This is especially true in the case here, where I was engaged in the two different disciplines of economic geography and international business management studies.

In order partly to prevent a descent into hopeless relativism, I imposed an order on this book which separates theory from empirical research, but it is not a reflection of the way the research itself was undertaken. The chronology implied in the structure, with theoretical reflection followed by field research followed by more theoretical reflection, is at best a gross simplification of the actual process of research.

The complexity of gaining access to those important actors for this research was a process that needed to be consolidated and reaffirmed throughout the research setting and research design, prior to the Indonesian fieldwork. Furthermore, I had to develop a two-way relationship in which the organisation can be seen to be receiving something in exchange for their time and trust. An article published in conference proceedings or an academic journal – the ultimate legitimisation – is not likely to be a sufficient return for the actors involved in this research, specifically from the MNEs' point of view.

In conducting fieldwork in Indonesia, organisational context is often experienced by the researcher as a direct and pragmatic concern, particularly at the stages of gaining access to the organisation and obtaining permission for in-depth data access. However, I believe that the implications of the organisational context are much wider. The nature of the actors involved (particularly the Japanese MNEs) affects research setting and research design, data collection, data interpretation, and analysis.

Even the research questions might be the result of a compromise between a researcher and the key contact person, and an outcome of a lengthy process of negotiation and renegotiation. Therefore, to overcome this issue, a researcher must carefully and reflexively consider relevant contextual factors when it comes to reliability when reporting the results of a study.

Researching the automotive industry with the involvement of MNEs is becoming more challenging as the MNEs are an organisational form experiencing rapid change. Not only is the organisational context becoming increasingly complex, but so is the external environment in which the organisation operates. The fact that MNEs are increasingly coming under pressure and scrutiny from global capital markets, thus compelling them to seek concrete pay-offs, is significant. Furthermore, in this dynamic environment, the influence of context can hardly be overstated.

In view of that, the relationships I developed were often personal ones, involving key supporters in the organisation. Access was therefore an ongoing process requiring me to invest considerable time and effort into communicating with key organisational members. In addition, I must tailor the presentation of the expected findings to the specific needs of the organisation, emphasise future implications rather than retrospective analysis, and be prepared to brainstorm key research ideas with organisational members.

The value of explanatory case study for this research is to increase knowledge about the 'real economy' in Indonesia due to global-local integrations in automotive industry production. Furthermore, this research aims to stimulate the debate about the strengths and the weaknesses of the Indonesian automotive cluster along with the importance of focusing on knowledge (pooling) and upgrading for the industry. Thus, by using case studies research, I can highlight the role of organisations in supporting and facilitating innovation and pave the way for more cluster-specific policymaking.

Reflection on the research process

In view of this, the research is a process of ongoing dialogues between me and the business world, an engaging with theory as well as articulating relationships and networks as part of the organisational fields, specifically for the automotive industry. In this final section I want to emphasise the importance of a qualitative approach.

Based on the Indonesian fieldwork during a four month intensive visit, I and my interviewees needed to establish an intersubjective understanding. In particular, the focused interview relied on the strength of eliciting answers whereby the interviewees used their own words and frame of reference. Therefore, the establishment of rapport in focused interview is of paramount importance given that the method itself is designated to elicit understanding of the interviewee's perspective.

In this research, the interviews were mostly semi-structured and accompanied by an ongoing email and phone correspondence both before and after the Indonesian fieldwork. Interview techniques were developed throughout the research experience, and enhanced rapport has been established over three and half years. Accordingly, in order to check the reliability of research techniques, interview results were cross-checked with other sources of data by way of mixture of pattern seeking and clustering of the sources of evidence.

Furthermore, the intention of the research is investigating the development of the automotive cluster in Indonesia, which has been regarded as a male-dominated workplace and interest, and mostly has to deal with a multicultural context. My past experience in dealing with the Japanese business environment was helpful in exploring and understanding the cultural values of the Japanese automotive industry and its business philosophy (i.e. *keiretsu* and Japanese business management).

The qualitative nature of this research was very helpful in incorporating this cultural dimension and has led to a deeper understanding of the development of the Java automotive cluster. For the kind of research conducted in this work, the qualitative approach was therefore a proper choice for the following reasons:

1   The explorative research questions ask for an in-depth collection of data and to let these data 'speak for themselves'. From there on, the connection to the theory can be made, instead of departing from a clear theoretical framework. The latter is more appropriate for explanatory research.
2   In connection to the previous point, this research did not start from clearly defined hypotheses. Instead, the why and how (i.e. the causal mechanism) behind the development have become clear over the course of the research. Qualitative research has provided the richness of data and the intimate understanding of the research subject through social contact and face-to-face relations in order to build trust between the interviewees and the researcher. That kind of relationship is necessary for this kind of research.

Data analysis

The next step is to analyse the data. This section explains the data analysis used in this research, starting from preliminary data to the qualitative approach after conducting fieldwork in Indonesia.

I used interview transcription in smooth verbatim format (and so I did not transcribe certain utterances); the content, rather than the exact wording, is of primary interest.

This subsection also highlights the importance of data validation for this research.

Qualitative data analysis

The primary data from research in Indonesia have been collected from the archive evidence, ongoing email correspondence, phone interviews with employees and other relevant representatives from the university, government bodies and industrial developers. In this regard, I started the analysis from the research questions in Chapter 1. For this reason, the analysis process has used the theories and the framework based on the literature review in Chapters 2 and 3, to identify the actual case in the Indonesian automotive cluster in Chapters 5 through to 9.

Throughout the data analysis process, I used a qualitative approach. The following paragraph elaborates the qualitative approach based on transcriptions of semi-structured interviews from Indonesian fieldwork and the analysis of secondary data (i.e. statistical and numerical data) collected from various respondents in documentary evidence analysis.

Qualitative research has always had a place in the field of international business and management (Marschan-Piekkari and Welch 2004). Early contributions on firms' internationalisation process and regional management typically drew upon a range of techniques, including qualitative methods in order to generate theory and new insights. In fact, qualitative research is often conflated with interview-based case studies and qualitative methods for data collection, and analysis may include participant observation, as well as unobtrusive methods such as archival research.

In view of that, I used semi-structured interviews for this research since questionnaires are often inadequate instruments for the complex issues of international business. Therefore, qualitative methods have been defined as a procedure for 'coming to terms with the meaning not the frequency' of a phenomenon by studying it in its social context (Van Maanen 1983; Werner 2002; White 2002).

The analysis of data here starts from the relevant theoretical frameworks, which have been explored in the literature review in Chapters 2 and 3. In conjunction with that, the empirical data from fieldwork in Indonesia has been critically analysed alongside the secondary data of Chapter 4.

I have used qualitative data analysis by linking the empirical and secondary data and the research questions. This is known as a 'pattern matching' process. Because of the flexibility of a case study (Yin 2003b; Bloomberg and Volpe 2008), it can be argued that a high-quality research case should demonstrate multiple sources of evidence; hence, the need for a mixture of pattern seeking and clustering of the sources of evidence, together with the ability to provide meaning, both of in-depth understanding of the central issue being explored, and a broad understanding of related issues and contexts.

Nevertheless, as case analysis is a difficult task, the researcher has used a mixture of pattern seeking and clustering. As a result, a researcher must bear in mind that the analysis of qualitative data with the use of case study as a research strategy is not an easy task: the construction of explanations needs to be done with rigour, care, and a great deal of intellectual and strategic thinking. As a result, the analysis of the empirical data, which is provided here in Chapters 5 to 10, was always likely to involve a lengthy period of time.

In fact, it took nearly ten months to comprehensively analyse the empirical data along with the secondary data after the Indonesian fieldwork, particularly in translating the smooth verbatim to thesis content. The thesis content is mostly used when the 'gist' of the recording is desired, perhaps for the purpose of summarising a discussion in the interview.

In order to produce a good standard of verbatim transcript, I listened to the audio many times to capture and to hear phrases and not simply type 'inaudible'. Accordingly, repeated listening was necessary to comprehend the meaning of the sentence. It took 'second passes' and another review to finally produce accurate interview transcripts. In the end, I formulated rough drafts.

In the context of analysing empirical data, I used cluster theory to focus on the linkages and interdependencies among the actors in the knowledge production chain in the Japanese automotive industry. Therefore, it is important to show the perspective from the Japanese subsidiaries in Indonesia alongside the perspective of the parent companies in Japan (the Toyota Motor Corporation and the Honda Motor Company). Interdependency is key to cluster analysis (Roelandt et al. 1999; Henderson 2002). However, the research on Indonesian automotive clusters has found that interdependency is multifaceted and is based on trade linkages, innovation linkages, and knowledge-flow linkages, or on a common knowledge base or common factor conditions.

Where cluster analysis has been undertaken, it has produced useful information on the actors involved in clusters, and on the value chain relations of firms. The dynamic interaction linkages have become evident as well as the institutional setting for cluster growth in the Java region. Accordingly, a clear distinction should be made between those approaches focusing on linkages between actors in networks or value chains and general quantitative cluster techniques (see Chapter 2 for the literature review on clusters) that aim to detect objects with similar characteristics.

In addition, interpreting and analysing qualitative data in this research has been perhaps the most difficult task. I cannot be satisfied merely with 'telling convincing stories' alongside the data analysis process. Therefore, in conducting qualitative analysis, *authenticity* is the main issue rather than reliability. The idea is to present an authentic understanding of a respondent's view.

Nevertheless, this means not just understanding the point of view of the individuals and the process being studied, but also that data must be interpreted against the background of the context in which they are produced. Ensuring data analysis and collection are closely interconnected during the life cycle of the case study research is a way to produce an authentic interpretation.

In view of the risks of generalisation in case study research, another way of tackling such criticism is to carefully select a case or cases that are typical and contain certain cluster characteristics which are representative of other cases (Maxwell 1996; Stake 2000; Marschan-Piekkari and Welch 2004). Therefore, the research setting was made followed by research design in choosing and composing the fieldwork. In other words, the principal use of the case study is not to test theoretical propositions but to find out the relationship of the case findings to theory. Case study research is about making analytical generalisations and not about making statistical inferences (Van Maanen 1983; Hoen 2002; Yin 2003a).

Additionally, the most frequently cited objection to the use of case study in social science research is the issue of representation, which raises the question of the extent to which the research findings can be generalised to a wider population beyond the case study (Yin 2003b). Consequently, I defend the research setting and research design in Chapter 5 as being a legitimate research strategy on philosophical grounds along with the key element for case study research in the research questions in Chapter 1.

Research limitations

Yet in the strength of this research also lie its limitations. The explorative nature of this study and the low number of observations make generalisation of the results problematic. Although the findings of this research will be relevant for other non-mature clusters in developing countries, this kind of research cannot verify or falsify hypotheses. But this research may yield testable hypotheses for

100 *Methodologies*

large network studies; for example, the hypotheses that underline the role of the leading MNEs to develop knowledge transfer in the automotive cluster.

Such a hypothesis can be tested by sending questionnaires to a large number of companies in several clusters in the automotive industry in other developing countries, asking them about their network relations, their knowledge transfer activities and their level of development. Although such an approach would not, as this research does, consider whole clusters, it would explain a vital causal mechanism behind the development of clusters.

Research ethics

Securing access to people, organisations, and data is necessary for successful completion of any research project (Yin 2003b). Without a doubt it is particularly important in this case study research where I have spent a considerable amount of time with a variety of relevant individuals and within a relatively boundless number of settings. Therefore, the issue of support should be considered in terms of negotiating data access (Burgess 1991; Creswell 1994), as during the research process it was inevitable that I came to know some individuals better than others. Accordingly, research networks through a snowball effect have been created during this research project. In this case, the snowball effect is defined as an approach for locating information-rich key informants, in which a few potential respondents are contacted and asked whether they know of anybody with the characteristics suitable for the research.

In term of ethical issues, on which the research focus is regarded as sensitive by management (particularly for Toyota and Honda and their suppliers), negotiation of access was in some cases a protracted process. Accordingly, I have used pseudonyms to protect the respondents' identities when subsequently reporting findings.

Furthermore, ethical issues can feel less immediate in this research as I have used documents and visual data to support the analysis. However, documents and visual data can take a very private or confidential form. Therefore, all the questions about ethical practice have been cross-checked with the respondents prior to the analysis of documents and visual data. The result of the analysis was returned to the respondents to verify the accuracy of interpretation.

Conclusion

Undertaking case studies for this research meant that the use of theoretical concepts to guide the design and data collection was an important strategy for completing successful case studies. The aim was to develop preliminary concepts at the beginning of a case study.

One purpose served by such concepts is to place the case study in the context of appropriate research literature, so that the lessons from the case study are more likely to advance knowledge and understanding of a given topic. A second purpose is to help define what the case is, in order to identify the criteria for selecting and screening potential data for the cases to be studied and to suggest the relevant variables of interest. Accordingly, without guidance from the preliminary theoretical concepts, all the choices might be difficult and impede the development of a rigorous case study.

I chose the explanatory case study as a research strategy after taking into consideration four key issues, namely:

1   Research questions
2   My skills and attitude
3   Costs or budget available
4   Time available and target date for PhD completion.

The explanatory case study in this research supplements the presence of explanatory theories on clusters and international business in the Japanese automotive industry, to furnish theoretical explanations with rich and extensive data collected through overseas fieldwork in Indonesia.

Each of the cases in this research (as explained in the research design) was investigated by reviewing pertinent documents, interviewing a wide array of individuals including actual or potential users of the research, and observing the actual research processes or mechanism. The case study protocol, therefore, is tightly geared towards explaining the theories based on the literature review in Chapters 2 and 3, and provides assurance that the diverse empirical data collection would involve converging lines of inquiry and analysis of the archive evidence.

Subsequently, a key aspect of the theories was their complexity for this research. This permitted pattern matching of a series of events as the main analytic process in each of the cases. In this research, the case study method relies on explanatory theories rather than on other methods, so that pattern matching in case study analysis permits case studies to investigate explanations in a single study of both Toyota and Honda.

In addition, the data collection process for this research was more complex than those used in other research strategies. In fact, the researcher must have a methodological versatility not necessarily required for using other strategies and must follow certain formal procedures to ensure quality control during the data collection process. Furthermore, the analysis of case study evidence is one of the least developed and most difficult aspects of doing case studies. Therefore, rigorous thinking along with the sufficient presentation of evidence and careful consideration of alternative interpretations are essential to analyse the data.

In this research, the explanatory case study has been chosen to undertake the research. The goal is to analyse the case study data by building up an explanation about the case, not necessarily to conclude a study but to develop ideas for further study.

In other words, the research purposes are specially suited for understanding the meaning in the Indonesian automotive cluster context, the process by which the case studies (i.e. Toyota and Honda) take place, and also for identifying unanticipated phenomena and influences in the Japanese automotive business in Asia. Although qualitative research is not restricted to this exploratory role, it is still an important strength of the case study method.

In conducting this research, I have interwoven data collection with data analysis right from the first case/interview. This allows theory to develop alongside the growing volume of data, allowing the research questions to be formulated and even reformulated at the same time. This often lead me to new questions and new subsequent data collection through ongoing correspondence with the respondents. It is often difficult for case study researchers to filter or discard irrelevant data before analysis commences. Therefore, early analysis on preliminary data prior to the empirical work reduced the risk that I might simply drown in the sheer volume of data.

In view of the arguments in favour of the case study method, while analysis might not form an isolated process, it does involve some distinct stages. Thus, in this research, the first step of analysis is to construct a case description and explanation by using a qualitative approach. This has helped me to understand 'how' things are developing and 'why' things occur the way they do.

In view of that, I must describe the phenomenon being studied and make complicated things understandable in their component parts. However, it can be hard to describe and explain something satisfactorily. Therefore, it is important to start with simple 'story telling' about a situation and progress in chronological order. Consequently, I have been able construct a map to locate different elements. This has finally led me to build a thorough understanding of an explanatory case study.

6 The importance of the Java region for the Indonesian automotive cluster

This research is aimed at studying an industrial cluster in the Java region, specifically in the automotive sector, in an integral way. The development of the automotive cluster in Java cannot be understood without a thorough insight into the characteristics of the region as a whole; hence, it is important to draw up a frame of analysis to take local contexts in the Java region into account and study their interrelations.

In view of that, cluster strategy in Indonesia has been explored in Chapter 4, along with the general economic background and policy context in Indonesia. Building on this, Chapter 6 explains the rationale for Java being the most favoured as well as the most important region for the automotive industry in Indonesia. Furthermore, Chapter 6 sketches a background for the analysis of the development of cluster strategy in Indonesia prior to the case studies presented in Chapters 7 and 8.

The importance of Java as an industrial location

Spatial concentration of economic activity occurs mainly because some regions have characteristics that are more attractive to firms than other regions. This implies that there are factors external to the microeconomics of firm-specific operations that boost productivity in one location compared to another (Venables 2003; Rosenthal and Strange 2003).

This leads to important questions: the first is descriptive and concerns the nature of these factors. The second addresses normative issues about the underlying processes that lead to external economies. Of particular interest are determinants of productivity enhancing localised factors that can be influenced by government policies in industry, trade, or science innovation. In Indonesia, the agglomeration of industry in Java has been accelerated by manufacturing investment by multinational enterprises (MNEs). The details are discussed in this section.

Manufacturing investment in Indonesia, which is mainly located in Java, is key for accelerating the growth of the cluster. Indonesia's manufacturing sector was formerly the most dynamic sector of the economy, and achieved a persistently high real growth of around 11 per cent p.a. during the 25 years preceding the Asian economic and financial crisis of 1997 (World Bank 2005). It was the growth rate of the manufacturing sector, however, that was most affected by the economic crisis.

The manufacturing sector contributed about 28 per cent of Indonesia's total gross domestic product (GDP) in 2001–2005, which was a slight decrease from the value of 29.7 per cent in 2000 (ADB 2005) due to the monetary crisis in Asia. This value shows that manufacturing has consistently been a significant contributor to GDP.

Nevertheless, the value of manufacturing investment in 2005 slightly dropped to −3 per cent from the value in 2001 due to political instability surrounding the presidential election of 2005 (MOSR 2002). Between 2000 and 2005 the sector contracted and expansion was forcibly curtailed. Furthermore, the proportion of the total Indonesian workforce employed in the manufacturing sector decreased from 13.3 per cent in 2001 to 12.3 per cent in 2005. The decline was attributed to the slight decrease in manufacturing investment during the same period (MOSR 2002; MOTIRI 2005).

The Indonesian government's main strategy for the future development of the manufacturing sector is to boost manufacturing investment in industries which have competitive and comparative advantages (e.g. the automotive industry). In the meantime, the government has continually focused on improving political, social, and economic conditions in order to create a friendly investment climate that will boost investment value in all sectors (MOSR 2002; MOTIRI 2005).

Apart from its geographical location, Java has become an important region for the following reasons:

1 Main industrial areas in Indonesia have always been situated overwhelmingly in Java, and most of Indonesian modern manufacturing establishments continue to be located in Java. The growth level of manufacturing firms established in Java has been increasing steadily in each year since 2001, from 81.39 per cent in 2001 and increasing to 81.71 per cent in 2003 and 2004. This compares with the growth of firms in manufacturing outside Java, which is on average between 18 and 19 per cent. This therefore describes an imbalance in the distribution of manufacturing among the regions in Indonesia.
2 Java's population comprises more than half of Indonesia's total population, and so offers a huge potential market in its own right.
3 Most investments, either foreign or domestic, have been concentrated in Java. During the period 2000–2005, around 63 per cent of total approved domestic investments were located in Java, while 66 per cent of total foreign direct investment (FDI) flowed to Java (BPS 2005).

As a result, the Indonesian government has been trying to balance the development of industry across the country, away from the massive concentration in Java, since lagging regional development has been an important component of the Indonesian industrial policy debate (MOSR 2002; ADB 2005).

Historically, Indonesia has had some explicit regional development policies. These have been motivated by economic and social disparities across the archipelago, and concerns about specific regions such as eastern Indonesia (Kuncoro 1996). Although eastern Indonesia contains several relatively affluent provinces endowed with natural resources, its development lags behind that of the western Indonesia region (Java and Sumatra), mainly due to low population density, the remoteness of some communities, inadequate infrastructure, and high transport costs.

Consequently, Indonesian regional policies have resorted to a range of measures (MOSR 2002). The central government in Jakarta has distributed investment grants focused on infrastructure and expanding regional human capital throughout the country (e.g. the pre-decentralisation Presidential Instruction scheme). Tax incentives and holidays have been promoted at both national and regional levels.

Nevertheless, their effectiveness for attracting FDI to eastern Indonesia has been low. Therefore, international agencies such as the World Bank are also undertaking transport improvement projects to bridge the development gap in the lagging eastern provinces. For instance, the World Bank for Indonesia Regions Transport Project (EIRTP) has covered 15 provinces in Indonesia's eastern islands, consisting of Kalimantan, Sulawesi, Bali, Nusa Tenggara, Maluku and Irian Barat, in an attempt to improve access to employment opportunities, health care, education, and other social services and facilities.

However, significant spatial inequalities in economic activity as well as in welfare are not unique to Indonesia. Such patterns tend largely to be the result of decisions by individual companies to locate in the main selected regions of Indonesia. This creates employment, generates incomes and triggers spillovers into rural, service, and supplying sectors. Understanding what motivates firm location decisions therefore helps explain the causes for spatial economic disparities.

In order to understand how firms select a location in Java, two factors have to be considered that influence their decision.

The first includes those reasons that are external to the firm. The Java region has a natural advantage that makes it relatively more attractive to different types of firms. This may include induced advantages such as the good transport infrastructure, which is the result of past public investment. These factors are central to the 'New Economic Geography' models, where firms tend to locate in areas that have high demand for their goods, facilitated through good transport infrastructure and thus market access (Krugman 1991a, 1991b; Fujita and Krugman 1995; Fujita *et al.* 1999).

Second, there are reasons for locating to the Java region that are more specific to the firm's production process and its interaction with suppliers, customers or competitors. These are *production externalities* where firms locate in proximity to other firms to benefit, for example, from knowledge or information transfers. Over a longer period of time, these factors may increase or decrease in importance. Initially, the conditions for the emergence of agglomeration economies might be due to natural endowment that historically encouraged early settlement and economic activity. These 'first nature' geographies (Venables 2003; Burgess and Venables 2004) take account of sheltered harbours, natural resource endowments, access to inputs, proximity to markets, and availability of basic infrastructure.

The initial benefits can trigger a self-reinforcing process that leads to the emergence of urban-industrial agglomerations in Java, to a point where the initial advantage responsible for the growth of the centre is no longer dominant. In view of that, Java and particularly West Java's Jabotabek[1] region has by far the highest concentration of manufacturing activity (Kuncoro 1996; Aswicahyono and Feridhanusetyawan 2004).

Historically, Java's fertile volcanic soils supported high population densities, and by the sixteenth century the port of Sunda Kelapa, in what is today Jakarta, had established itself as an important trade hub (Hill 1997). This in turn attracted the establishment of European trading posts and eventually formed the capital of the Republic of Indonesia. In the post-colonial period, Indonesia developed what some have called an economic system of 'bureaucratic capitalism', often created by the state, by members of the government or military and their associates, or by ethnic Chinese businessmen (Ardani 1992; Tambunan 2005).

There was no prominent class of indigenous entrepreneurs, which might have created a more dispersed pattern of industrial development (Hill 1997; Tambunan 2005). These factors accelerated the agglomeration of economic activity near the seat of power in a highly centralised political system and resulted in the rapid growth of the manufacturing sector in the Jakarta region in the 1980s and early 1990s. By 1991 the manufacturing share of GDP in Indonesia exceeded that of agriculture and most of that was generated in West Java (MOSR 2002).

Additional concentrations of manufacturing include Surabaya in East Java – originally a Dutch centre of naval industry – and Bandung in West Java at the centre of the highland plantation economy and recently a centre of textile manufacturing. Two smaller manufacturing centres, Medan in Sumatra Island and Makassar in Southern Sulawesi, also owed their existence largely to their role as regional trading posts (Kuncoro 1996, 2002).

These agglomerations are supported by public infrastructure, particularly in the transport sector. Krugman (1991b) shows that manufacturing firms tend to locate in regions with larger market demand to realise scale economies and minimise transportation costs. If transport costs are very high, then activity is dispersed. In an extreme scenario, every location must produce everything locally. If transport costs are unreliable, firms may be randomly distributed as proximity to markets or suppliers will not matter. Therefore, agglomeration occurs where there are intermediate transport costs particularly when the spatial mobility of labour is low (Fujita and Thisse 1996). Low transport costs allow larger scale production, which in turn creates investment activities in other industries. By increasing demand for each other's products, this process of simultaneous investment raises profitability thus allowing all firms to realise pecuniary benefits (Murphy *et al.* 1989).

In addition to the pure pecuniary benefits that arise from reduced transport costs, the availability of good infrastructure increases the potential for input diversity, a larger labour pool, as well as the probability of technology diffusion through interaction and knowledge spillovers between firms (Henderson et al. 2001). Thus, improved accessibility has the effect of reducing geographic barriers to interaction, which increases specialised labour supply and facilitates information exchange, technology diffusion, and other beneficial spillovers that have a self-reinforcing effect (Lall et al. 2004; McCann 1998).

Other region specific characteristics that influence industry location decisions include local labour costs and administrative policies that support or obstruct business development such as regulation, taxation and amenity provision. A range of registrations, operating licences (e.g. zoning permits, environmental impact assessments), minimum wage and employment regulations, and taxation can influence the cost of doing business in a particular location. This also includes the prevalence of informal payments, outright corruption, or 'predatory' taxation. Firms in Indonesia must interact with the bureaucracy at the national (Jakarta) and regional level (i.e. provincial and local governments)[2] (Perdana and Friawan 2007).

While the central bureaucracy may not formally differentiate in its licensing processes across the regions, transaction costs may increase for firms located further from the central bureaucracy (e.g. owing to delays and uncertainty stemming from absence of face-to-face contact) (Perdana and Friawan 2007). At the regional level, certain provinces of local bureaucracies may simply be more prone to corruption and/or predatory taxation. At the same time, accountability mechanisms (e.g. an exit option on the part of firms) may not be sufficient to act as a disciplinary device on local bureaucracies. Since Indonesia's decentralisation in 2001, businesses have primarily complained about the nuisance caused or even predatory taxation by local governments.

Production externalities, in contrast, relate to dynamics that directly affect the firm's microeconomic decision-making (Szulanski 2000). Most fundamentally, clusters of firms that are predominantly in the same sector take advantage of local economies. They include sharing of sector specific inputs, skilled labour and knowledge, intra-industry linkages, and opportunities for efficient subcontracting. Marshall–Arrow–Romer[3] externalities suggest that cost-saving externalities are maximised when a local industry is specialised. These models predict that these externalities predominantly occur within the same industry. Therefore, if an industry is subject to such externalities, firms are likely to locate in regions where other producers of that industry are already clustered.

The economic, spatial and cultural context in Java

In general, the sector-mix in the urban region of Java influences the development of the specific growth cluster (i.e. automotive cluster) under consideration, because there are often many interrelations (supplier-links, demand, services) between the growth sector and another sector in the Java region.

Another important characteristic of the urban region is the knowledge base of the local economy as reflected in the education level of the workforce, the knowledge-intensity of the economic activities that take place, and the presence of research institutes or universities. The assumption is that in most new growth clusters, knowledge is the key driving force behind growth and development. Therefore, regions with a well-educated workforce and high-quality knowledge institutes will have a general advantage over the other regions. In this case, the Java region is more developed and advanced in terms of education and knowledge-intensity in contrast to other regions in Indonesia.

The spatial conditions in the Java area shape the second context element of the analysis of this research, which has two main concerns.

The first is the quality of life in the urban area of Java. Generally, quality of life is one of the most important factors in the choice of location (Breziz and Krugman 1993). In the automotive cluster case, firms move to areas where they can find appropriately skilled people, attached to a high standard of living environment, such as the Java region (Sonis et al. 1997; Kuncoro 2002; Tarmidi 2004).

The second concern is accessibility, which can be split into physical and electronic, or internal and external to the urban region. Accessibility for firms in the Java region is a necessary condition for manufacturing development, since interaction is crucial in a network economy. Bad transport systems, particularly in the automotive industry, may significantly impede interaction in a cluster, particularly if cluster elements are dispersed. Moreover, external accessibility to other cities and regions is fundamental, as a means to link local networks up with national and international networks of all kinds. Simultaneously, the level of external accessibility influences competition between cities for the development of the cluster.

In addition, the 'cultware' as mentioned by Van Den Berg et al. (2001a, 2001b) is considered an element of the cluster context as an institutional variable. Cultware relates to attitudes of people and firms. In particular, attitudes towards innovation are important in the automotive sector as the main driving force of the development of the cluster, alongside the willingness of actors in the cluster to cooperate. Cooperation is also one of the main sources of development, new combinations, and growth of the cluster (Nooteboom 2003; Gallaud and Torre 2004).

Indonesia's manufacturing industry is showing real improvement since 2005 as macro- and microeconomic indicators are at an all-time high, confidence of business and consumers towards the government's economic team is greatly improving, and the public hopes that a better future is imminent (MOSR 2002).

For a start, all macroeconomic indicators remain positive, and are still improving: 2007 started off with GDP growth at 5.6 per cent while year-to-date inflation was 5.32 per cent. The interest rate was lowered to 9 per cent by the end of first quarter, which was good news for the manufacturing sector; hence, Indonesia's export manufacturing industry growth is expected to be steady (MOTIRI 2005).

Nevertheless, poor infrastructure, legal uncertainty and a lack of tax incentives have been blamed for declining foreign investment in the automotive sector. Still it remains a key pillar of the economy with investments totalling more than US$7 billion and generating employment of over 300,000 (MOTIRI 2005). The economy expanded by 5.1 per cent in 2007, mainly spurred by consumption, and growth is expected to continue, spurring hopes of continued demand for the automotive sector.

Automotive cluster characteristics in Java

The important components of a cluster's functioning are its size and its development level (i.e. whether or not the cluster possesses critical mass), and how many companies and educational and research institutions are active in the cluster.

There are several reasons why critical mass is important and leads to an elaboration of the automotive cluster's characteristics in Java. As can be seen in Figure 6.1, the interrelatedness of internal and external networks is an important element for explaining the characteristics of the automotive cluster in Java. The discussion is elaborated in this section.

First, critical mass ensures a market large enough to support (specialist) activities in the cluster. In May 2007, Indonesia's vehicle market remained on track for strong recovery from the major slump in 2006. PT Toyota-Astra Motor, the leading local vehicle distributor, estimated an increase in sales of 66 per cent to 38,313 units from 23,079 in May 2006. The initial five months of 2006 registered cumulative sales of 158,189 units, a 26.5 per cent increase over 125,068 units in the year before. High optimism rules the industry given that the full-year prospects and expectations for sales were for 380,000–390,000 units in 2007, in contrast to 317,000 units the previous year (TMI 2007).

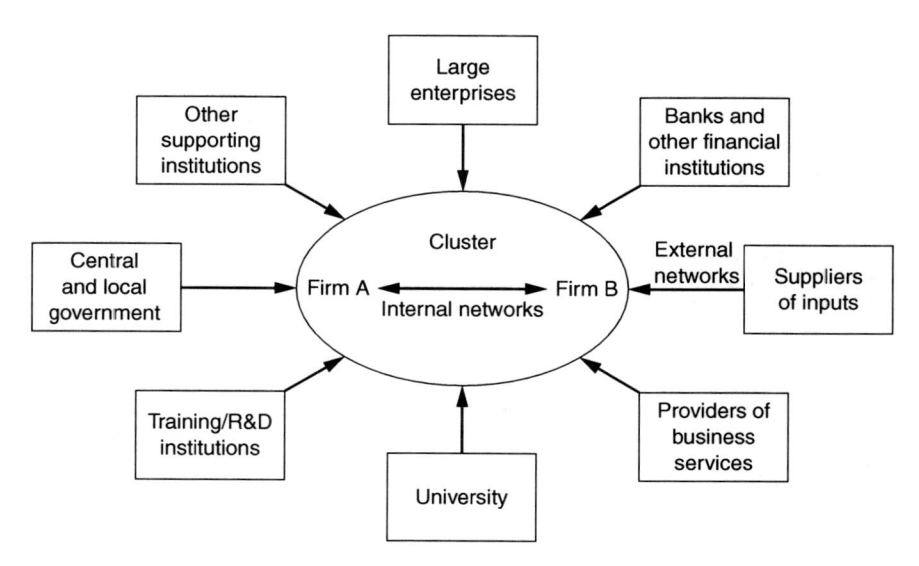

Figure 6.1 The automotive cluster in Java, Indonesia (source: author's fieldwork (2007)).

110 *The importance of the Java region*

Additionally, the Ministry of Industry also forecast that automotive exports in completely built-up (CBU) and completely knock-down (CKD) condition may increase in the next few years. Exports of CBU units in 2007 was 34,627, increasing from 30,307 in 2006, while exports of CKD units had gone up from 105,022 in 2006 to 118,045 in 2007. Moreover, the Ministry of Industry also projected an increase in investment in the automotive industry from about Rp3.9 trillion to Rp4.1 trillion in 2007, boosting production capacity utilisation in the national automotive industry from 38 per cent to 43 per cent.

These projections were proven by a 2007 announcement by one of the global automotive giants, Toyota, which expressed its plan to designate Indonesia as one of its main production bases for the supply of Multi Purpose Vehicles (MPV) to the global market. With planned exports of Toyota Fortuner and Avanza to the Middle East, Latin America, Africa and Asia, Toyota Indonesia will overtake Toyota Motor Thailand as the largest supplier of Fortuner cars to the Middle East market. Carmakers are also engaged in expansion of their current production capacity to keep pace with demand and take advantage of the market potential due to the fairly low car ownership ratio in a country that has the fourth largest population in the world.

Second, the presence of many companies may encourage keen competition and thus push the companies to operate in more effective and efficient ways. In the automotive cluster in Java, manufacturing activities have spread across different locations in different industrial estates, although some concentrations have taken place in some main industrial estates. Table 6.1 provides details of the major car cluster locations in Java.[4]

Table 6.1 shows that currently there are 16 major automotive firms in Indonesia, which together produce 22 brands of automobile. Total capacity of

Table 6.1 Major carmakers, Java[4]

No.	Assemblers/manufacturers	Location
1	Volkswagen	DKI Jakarta
2	Isuzu Motor	DKI Jakarta
3	BMW Indonesia	DKI Jakarta
4	Astra Daihatsu Motor	DKI Jakarta
5	Peugeot	DKI Jakarta
6	Mazda National Motor	DKI Jakarta
7	Mitsubishi Krama Yudha	Bekasi
8	Hyundai Motor Indonesia	Bekasi
9	General Motors	Bekasi
10	Indomobil Suzuki International	Bekasi
11	Toyota Motor Manufacturing Indonesia	Karawang-West Java
12	Honda Prospect Motor	Karawang-West Java
13	KIA motor and Timor Motor	Karawang-West Java
14	Nissan Motor	Karawang-West Java
15	Hino Mobil	Karawang-West Java
16	Mercedes Benz	Bogor-West Java

Source: author's fieldwork (2007).

the assembling manufacturers is more than 800,000 units per year. In Indonesia, more than 170 companies are producing various types of automotive components, covering 36 six-digit HS (Harmonized System) numbers.

Table 6.2 shows that nine major players in the motorcycle industry in the Java region are dominated by the Japanese.[5] Locations of carmakers and motorcycle makers are shown in Map 6.1.

Third, the chance of fast penetration of all types of innovation becomes greater when the cluster is larger. In studying the automotive industry, inevitably the technology is characterised by a specific knowledge base (Rasiah *et al.* 2008). In the automotive sector innovation is quite rapid, and sector boundaries change over time. In fact, knowledge and basic technologies constitute major constraints on the complexity of the automotive sector. Links and complementarities among artefacts and activities also play a major role in defining the real boundaries of a sectoral system.

In ASEAN-4 clusters, particularly Indonesia, knowledge transfer from Japanese car producers has played a crucial role for developing knowledge and technology in the automotive industry (Chen 1996). Nowhere in the world is the influence of transferred Japanese technology greater than in the ASEAN region (Chen 1996: 7). For Japan, this region has always been a crucially important source of raw materials and cheap labour for its own dynamic manufacturing industry. Furthermore, Indonesia has become not only a springboard for Japanese products to Western European and North American markets, but is also itself one of the fastest growing markets for Japanese products. Japan's technology transfer to Indonesia has been designed to strengthen and develop its ties with this country.

Additionally, the Indonesian automotive industry to some extent has benefited from technology and innovation including process and production technology. Knowledge accumulation has been embedded in this cluster when MNEs decided to develop and enhance production know-how, by branching out from assembly to several other complicated production processes, namely, the creation of stamping plants, casting plants, welding plants, and engine plants.

Table 6.2 Major motorcycle makers, Java

No.	Motorcycle company	Location
1	Astra Honda Motor	DKI Jakarta and Bekasi
2	Yamaha Indonesia Motor MFG	DKI Jakarta
3	Danmotors Vespa Indonesia	DKI Jakarta
4	Kawasaki Motor Indonesia	DKI Jakarta
5	Indomobil Suzuki International	Bekasi
6	Kymco Motor	Bekasi
7	Yamaha Motor Indonesia	Karawang-West Java
8	TVS	Karawang-West Java
9	Sanex Motor International	Tangerang-DKI Jakarta

Source: author's fieldwork (2007).

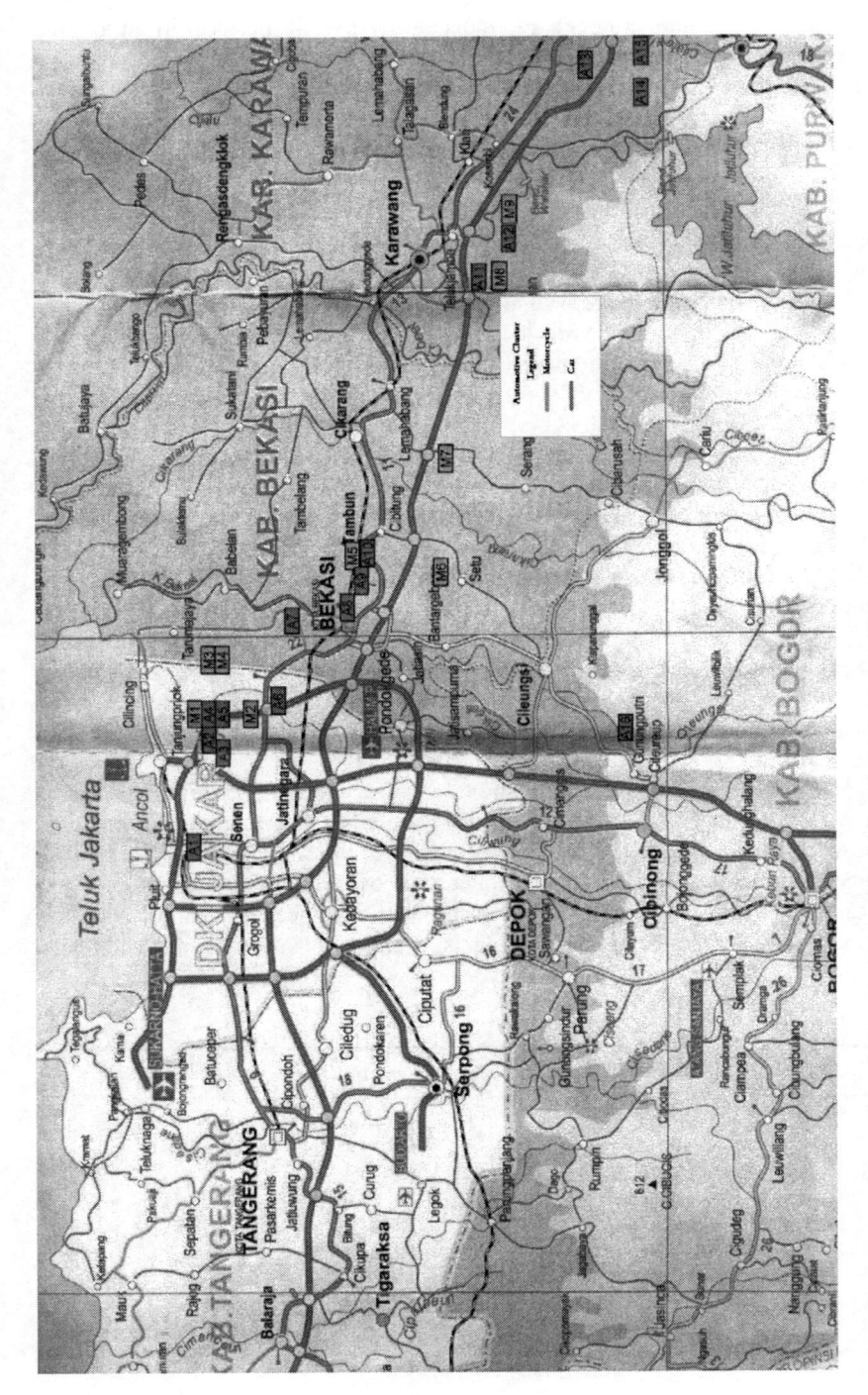

Map 6.1 The automotive cluster in Java, Indonesia (source: author's fieldwork (2007)).

Therefore, the process of knowledge transfer from an MNE to its local plants and its suppliers has been embedded in the automotive cluster in Java since 1967.

Fourth, regional cooperation is easier to accomplish within a large cluster, as it is easier to find a complementary partner or actor in the region (Steiner 1998; Kuncoro 2002; Martin and Sunley 2006). In fact, scale offers more prospects for the sharing of resources, the benefits of a shared pool of specialised labour, and scope for cluster benefits such as joint education facilities. In conjunction with the role of actors, the quality of actors will lead to the quality of the cluster. Quality refers to the degree of international competitiveness of firms, the technological sophistication of their output, the standing of a university, etc. Therefore, the presence of one or more engines in a region – either large multinational firms or other actors – is supposed to be a determinant of a cluster's functioning, in their role of expanding the network in global and local contexts.

In the automotive industry, heterogeneous agents are connected in various ways through market and non-market relationships (Fujimoto 2007). The types and structures of relationships and networks from the automotive sector is complex, as a consequence of the features of the knowledge base, the relevant learning processes, the basic technology, the characteristic of demand, the key links, and the dynamic complementarities.

In the automotive industry in Indonesia, Japanese corporations are more internally decentralised and more open to the surrounding economy than their US and European counterparts (Chen 1996; Rasiah 2005; Wee 2005). Thus, Japanese automotive producers rely on extensive networks of small- and medium-sized suppliers (i.e. Tier 1 and Tier 2), to which they are linked through ties of trust and partial ownership. Although Japanese producers historically exploited suppliers, many increasingly collaborate with them, encouraging them to expand their technological capabilities and organisational autonomy (i.e. the 'Toyota Way'). Therefore, these producers tend to be geographically clustered and depend heavily on informal information exchange as well as more formal forms of cooperation (Tarmidi 2004; MOTIRI 2005).

Furthermore, there are large as well as small firm variants of network-based systems. Large corporations can integrate into regional networks through a process of internal decentralisation. As newly independent business units are forced by competition to achieve the technical and productive standards of outsiders, they often draw on the social and technical infrastructure of the local economy and collaborate with external suppliers and customers.

As Saxenian (1996) points out, economic activity does not cluster within a single regional economy, particularly the automotive industry, where firms in network systems serve global markets and collaborate with distant customers, suppliers, and competitors. Thus, the automotive industry is highly international. However, the most strategic relationships are often local because of the importance of timeliness and face-to-face communication for rapid product development. Moreover, non-local suppliers succeed in part by integrating into regional economies that specialise in similar lines of business (e.g. original equipment component (OEM), replacement parts manufacturing, replacement parts distribution, and rubber fabricating). Paradoxically, the creation of regional clusters and the globalisation of production go hand in hand, as firms reinforce the dynamism of their own localities by linking them to similar regional automotive clusters elsewhere.

Along with research-academic partnerships, automotive makers in Indonesia have been in contact with local educational and research institutions to develop partnerships for developing particular engineering tests and knowledge-sharing activities, by providing on the job training programmes for particular engineering students to have an apprentice programme in some automotive plants (Perdana and Friawan 2007).

Furthermore, Toyota has developed a vocational engineering school to educate professionals in the automotive sector and enhance the practical engineering knowledge of employees in the Toyota group in Indonesia. Toyota has also built a learning centre for intensive knowledge-sharing among the Toyota group and then appointed a leader project to share with the non-Toyota group in the automotive cluster across the Java region. The employee exchange programme between Indonesia and Japan has also been accelerating knowledge exchange between the parent company and its Indonesian subsidiaries. Details of the case study are available in Chapter 7.

In Honda's case, the company has built a learning centre for sharing and developing ideas in automotive engineering for both suppliers and chosen universities. In addition, an interchange employee programme between Indonesia and Japan has been taking place since the 1990s. Details of the case study are available in Chapter 8.

Organising capacity of the automotive cluster in Java

The final element to play an important role in the performance of the cluster is its degree of organising capacity. Van Den Berg *et al.* (1999) mention that organising capacity can be defined as the ability of the urban region to enlist all actors involved in the growth cluster, and with their help generate new ideas and develop and implement policy designed to respond to developments and create conditions for sustainable development of the cluster.

Organising capacity can refer to the development of cluster-specific policy, the attraction of cluster-supporting elements (companies), investment in specific infrastructure, etc. Van Den Berg *et al.* (1999) differentiate some necessary elements for the organising capacity in general: leadership, vision/strategy, political/societal support, and university–industry–government partnership. All these elements are significant for the development of a cluster in a city or region. As a result, the development of the cluster should be steered by an integral vision, and preferably be underpinned as a strategy. Therefore, a well-defined strategy on the development possibilities of a cluster is indispensable for an efficient allocation of resources and efforts to stimulate the cluster.

In conjunction with organising capacity, political and societal support are also necessary conditions for a cluster policy (Roelandt et al. 1999; Benneworth and Charles 2001). Political support helps to bring about positive collaboration at a local level. In fact, proper presentation and communication of policies are of principal importance to achieve results. Societal support is important for the acceptance of policies aimed at growth clusters.

Additionally, university–industry–government partnership on strategic, tactical, and operational levels is very important for a successful cluster policy (Etzkowitz et al. 2007). An essential factor for success is the early involvement of the private sector in the development of locations, the attraction of companies, etc. The knowledge, expertise, and involvement of the private sector can be very valuable to the decision-making process, and considerably enhance the chance of success. Moreover, the government can act as a network broker, stimulating the formation of inter- and intra-sectoral networks, by bringing people and firms together (Porter 1998b; Perdana and Friawan 2007). Local or regional government can engage in university–industry–government partnership directed at the stimulation of the growth cluster, for instance by providing facilities or specific education.

In the early stages of Indonesian economic development, much of the investment came from the government, while the contribution of the private sector was still relatively small. This was because in that period, the government had relatively big savings, especially from oil/gas receipts, which enabled it to put in substantial investment. On the other hand, private enterprises had not developed due to, among other things, lack of capital and infrastructure.

Afterwards, as development funds from the government became more limited, the government concentrated its investment in infrastructure development, which stimulated and strengthened the private sector, working on the assumption that the private sector would play a greater role in economic development; hence, stronger private capital brought a change in the composition of investment funding.

Following the issue of Law No. 1 of 1967 on Foreign Investment (PMA), which was amended by Law No. 11 of 1970, and Law No. 6 of 1968 on Domestic Investment (PMDN), which was amended by Law No. 12 of 1970, several policies were issued to improve the investment climate in Indonesia. It is widely recognised that the impressive increase in private investment had something to do with the government's efforts to create an encouraging investment climate for both domestic and foreign investors.

With regard to government cluster strategy for manufacturing in Indonesia, the automotive sector had first agglomerated organically long before the cluster strategy was formed. The auto firms' decision to be located in the non-industrial estate areas in the north of Jakarta (i.e. Sunter) has organically grown since the 1970s (Kuncoro 1996, 2002; Aswicahyono et al. 1999).

Nevertheless, since 1989, when the government issued a policy on industrial estates to provide better infrastructure for manufacturing sectors, the procedure for investment has changed (Aswicahyono et al. 1999; Tarmidi 2004, HKI

2006). Now, the government will suggest that the investor's manufacturing investment should be located within certain industrial estates rather than outside them.

Under this new regulation, since 1989 the automotive industry has had to locate their manufacturing plants in selected industrial estates across Java. Therefore, most of the automotive industries have deliberately chosen the Jabotabek area for their main manufacturing bases. As a result, the cluster growth for the automotive industry in Java has formed networks with different actors involved.

The actors involved are: the government bodies in seven relevant ministries (i.e. Ministry of Industry, Ministry of Manpower and Transmigration, Ministry of Trade, Ministry of Transportation, Ministry of Education, Ministry of Information and Telecommunication, and Ministry of Research and Technology), the learning forum together with a working group and cluster facilitator (i.e. industrial estate developer), the local/provincial government with local institutions, the universities and R&D institutions, services firms, and the automotive association. The internal network includes *jishuken*, a competitive learning and working group in the Japanese *keiretsu*, while both central and local government, universities and R&D institutions, service industries, and the automotive associations form the external network.

Starting from raw material production to machining and engine assembling, followed by marketing processes, the ministries are nurturing the raw material stage by providing tax incentives and policies to support the gathering of raw material from domestic and international suppliers. The universities and R&D institutions are helping the automotive industry in material testing and early engineering stages. The learning forum and working group (i.e. *jishuken*) enables engineers to exchange knowledge in the production system. This stage is important as the engine and body are produced and assembled in detail. Also important is the role of service industries, for example, banks, transportation/delivery services, and insurances. The last stage is the marketing and distribution in which local government and the automotive association are involved.

Accordingly, this interrelatedness of internal and external networks has supported the growth of the automotive cluster in Java. In fact, the changed bureaucracy has enhanced cluster growth in the automotive sector, from organic growth to 'artificial' growth, which makes it easier for the internal and external network within and outside of the cluster to pursue collaboration (HKI 2006; Perdana and Friawan 2007). In fact, in the case of the car and motorcycle industries, most of the car- and motorcycle makers and their suppliers have deliberately chosen to be located in the industrial estates across Java in order to support their production, business network, and knowledge transfer (Tarmidi 2004; HKI 2006; TMI 2007; Honda 2007a, 2007b). Accordingly, this interrelatedness of internal and external networks has supported the growth of the automotive cluster in Java.

Additionally, the Indonesian government has set up a strategic plan to manage the ongoing process in the automotive cluster. The automotive sector consists of car and motorcycle production, which triggers the establishment of supportive industries and related service industries. The plan has been divided into the short term (2004–2009) and long term (2010–2025), each of which has different operational actions.

The government strategic plan has been developed based on the characteristics of the Indonesian automotive industry (MOTIRI 2005) which has at least six important characteristics, most of them with universally applicability. They are explained as follows:

1 It is a scale-intensive industry, particularly in some major manufacturing activities, for instance casting and engine construction.
2 It is generally technology- and capital-intensive.
3 There is pronounced product differentiation, as illustrated by the importance of brand-name recognition in this industry.
4 The industry comprises a diverse collection of interrelated subsectors, of greatly varying technological, capital and scale intensity.
5 In consequence particularly of the first three factors, the industry is almost completely dominated by MNEs; globally, it is one of the most highly concentrated industries in the world.
6 The industry invariably attracts more government intervention than almost any other manufacturing activity, particularly in the Indonesian context. It is seen as a vehicle for the introduction and diffusion of new technologies – in particular a means of technology transfer from MNEs – and as an instrument for the development of small- and medium-sized enterprises (SMEs) through subcontracting arrangements.

Operational actions are managed under the central government (divided into four ministries), local government (i.e. provincial and district), and private sectors (association and industry). Central government implements all the operational actions in strengthening cluster growth, while the local government supports firms mainly through tax and credit incentives (Aswicahyono et al. 1999).

The role of university and R&D institutions are less significant as they support the production process along with quality improvement. Additionally, the role of knowledge transfer and capacity building division is quite interesting as it is divided into three key roles: the competitive learning forum, the working group, and the cluster facilitator. The learning forum here is the automotive association in the automotive industry, which mainly emphasises technology, marketing, and investment. The working group forum is the engineer learning group within the automotive industry whereby they can share knowledge of the production system and of the technological aspects. Therefore, this is much more focused on developing knowledge and technology. Finally, the cluster facilitator (i.e. industrial developer) is the intermediary between the automotive industry, government, university, and another business network, to expand the internal and external networks within and outside the cluster.

Conclusion

It can be argued that Java has received more policy and analytical attention than other regions in Indonesia. It was one of the first industrial regions to be established in Indonesia from the late 1920s and has been the subject of intense policy intervention, variously regarded as a 'spearhead' for technological modernisation and a vehicle for the diffusion of imported know-how (principally from Japanese FDI) in Indonesia.

Subsequently, improving human resource management specifically in production engineering as well as business management is still needed to boost Indonesian industry. Therefore, the Indonesian government has set up five priorities to support the industrial cluster as a prime mover in the economy. The framework used to set up priority industry is based on international competitive advantage and the Indonesian potential factor in developing industrial conditions. As a result, the Java region is deliberately considered the main location for the automotive industry in Indonesia.

The Indonesian car industry is heading for a strong recovery from the slump in 2005. The car industry is also engaged in expansion of its current production capacity to keep pace with demand and to take advantage of the market potential due to the fairly high car ownership ratio in the country, which has the fourth largest population in the world. In parallel, looking at the appropriateness of the motorcycle as a cheap alternative to personal transportation for developing countries, this is also an important element in the automotive industry.

Nevertheless, the industry's development has not been trouble-free as it involves different actors from both internal and external networks of the automotive cluster. This means that more collaboration with the relevant institution is needed to develop infrastructure in those areas where existing or potential industry is likely to take place.

In addition, the Indonesian government still needs to enhance the relevant industrial policy in order to support the existence of manufacturing industry. There needs to be better synchronisation between industrial policy, investment, and trading policy to be conducive to business opportunities in Indonesia. Also, paring down the bureaucratic process for MNEs in Indonesia is highly necessary. In this book, the relevant policy analysis and recommendations are elaborated in Chapter 9.

7 Car production in Indonesia
The Toyota complex

Chapter 7 discusses the Toyota case study in order to understand the impact of Toyota on automotive cluster transplants, through knowledge transfer as well as through a localisation project within the cluster. In the case of Toyota, a strong finding is the role of the knowledge transfer process between Toyota Japan and Toyota Indonesia suppliers through *jishuken* and TPS (Toyota Production System). However, this chapter also discusses the importance of transferring knowledge between Toyota Indonesia and its supplier in the automotive cluster.

Chapter 7 provides a case study of Toyota in Indonesia in answer to two of the research questions posed in Chapter 1:

1 **Outcomes:** How has Toyota contributed to the development of the Indonesian automotive cluster?
2 **Process:** How has Toyota been fostering technological change and knowledge-technology transfer in the Indonesian automotive cluster (through localisation project and suppliers)?

The case study also looks at the Kijang Project as a localisation project from Toyota Indonesia, and the relation between Toyota Indonesia and a local academic institution in developing local knowledge in the automotive sector.

In addition, Chapter 7 highlights the place of the Indonesian subsidiary within the Japanese automotive multinational enterprise (MNE). It begins with an explanation of Toyota's internationalisation strategy and the establishment of Toyota Indonesia. Moreover, this chapter gives emphasis to the way Toyota's development both in management and technology is affected by being located in the automotive cluster in the Java region.

The Indonesian subsidiary within the Japanese MNE

Toyota's internationalisation strategy

As an international player in the car-manufacturing industry, Toyota has decided to expand its integrated production network globally, mainly in Asia as its Pax-Niponica (Gerlach 1992; Han 1994; Dyer and Nobeoka 2000). As a result, it is important for Toyota to penetrate overseas markets as a production base, particularly in ASEAN-4 automotive clusters such as Indonesia.

Regarding overseas plant production, Toyota has underpinned this initiative as part of its mass production spatial strategy. This means that the product development costs of a single model could be offset over a greater production run if the car was designed to sell in all lucrative markets (potentially outside Japan) but produced 'locally' (Dohse *et al.* 1985; Liker and Meier 2005; Hino 2005). It also takes into account the basic standardised Toyota Production System to maximise lean production through knowledge transfer (Liker 2004a, 2004b; Asheim and Coenen 2006). Therefore, in order to establish a lean production system it is significant to have Just in Time (JIT) delivery systems alongside *kanban* as it requires many small loads of parts to be delivered frequently to assembly plants, which accelerates the growth of local firms in the supply chain (Sugiyama and Fujimoto 2000; Fujimoto 2007). Accordingly, to bring together the entire range of Japanese production characteristics, it is important to include the role of Japanese automotive networks in their *keiretsu* (Kenney and Florida 1993; Dyer and Nobeoka 2000; Fujimoto 2007). For that reason, Toyota has expanded its existing production capacity to meet demand and exploit market potential, suggesting potential for even more growth.

On the other hand, the Indonesian automotive cluster is increasingly spread across international boundaries and Indonesia has been trying hard to get back on the right track (Aswicahyono *et al.* 1999). The automotive industry in general has accelerated economic transformation, particularly in the manufacturing sector, since the 1970s. Since then, the automotive industry in Indonesia has become a national status symbol, impelling Indonesia to launch export-based production by developing an ongoing learning process in automotive engineering. The role of clusters as the location for automotive MNEs has been crucial in this context (Tarmidi 2004; Wee 2005; MOTIRI 2005).

Establishment of Toyota Indonesia in the automotive cluster

Toyota Indonesia was established in 1970 as PT Toyota Motor Manufacturing Indonesia (PT TMMIN) with a total number of 5,143 employees up to 2007. Additionally, in 2004, Toyota Indonesia established Toyota-Astra Daihatsu Motor to produce the Daihatsu brand as part of Toyota *keiretsu*.

In 1973, PT TMMIN established the first assembly plant in an automotive cluster in Sunter (Jakarta). The Sunter plants cover 310,898 square metres of land, with a building covering 175,986 square metres. This site was the first Toyota plant in Indonesia, which modifies modern technology with human skill resources.

The Sunter plants consist of five plants with different functions: casting, stamping, engine, packing and vanning, and waste water treatment.

Each of the five plants in the Sunter base is described here. The casting plant is where machining components are produced. It covers a 65,028 square metre area with a production volume of 1,000 tonnes per month with two shifts in the operational schedule. This casting plant is provided with facilities for an induction holding furnace, sand blasting, sand mixer, sand reclaimer, vacuum sand conveyor, and drying oven.

Second is the stamping plant. It has a production capacity of up to 96,000 units per year, with an area of 64,247 square metres. This plant is specifically intended to produce press parts for Kijang Innova, Kijang Avanza and Dyna/Hino. This plant has been provided with press machines, milling and drilling machines, die spotting machines, measuring machines, trial machines, and pressing machines.

Third is the engine plant intended for engine 7K. It has a production capacity of 4,400 units/month with an area of 15,327 square metres. The engine plant produces machine type 7K and 14B for the Kijang Pick Up and the Dyna Truck. Additionally, this plant is responsible for producing machine components, assembling machine 7K, assembling and packing for machine type TR and its components for export to Thailand, Venezuela, South Africa, and the Philippines. For Engine TR, with an area of 19,000 square metres, it has a production capacity of up to 15,000 units per month on the machining line and 13,000 units per month on the assembly line. The Engine TR plant has been equipped with facilities for machining lines, assembly lines, interlock systems, supply part systems, leak testers and test benches.

Fourth is the packing and vanning plant. The packing for exports is done in this plant of 7,200 square metres. The capacity of the packing plant is up to 4,200 units per month for Kijang Avanza component and 5,000 units per month for Kijang Innova components. Those completely knocked-down (CKD) units are sent to the Philippines, Malaysia, Vietnam, Argentina, South Africa, Venezuela and Brazil; while the CKD for passenger vehicles are sent to Thailand, India, Vietnam, Taiwan, and South Africa.

In 1998 Toyota Indonesia expanded its manufacturing and assembling plants outside the automotive cluster in Sunter-Jakarta. The chosen location was the industrial estate in West Java, Karawang International Industrial City (KIIC), which later generated growth of another automotive cluster (including the auto component industry).

The Karawang plant has been operating since 2000 with a total investment of Rp462.2 million. This plant has a land area of 1,000,000 square metres with a 300,000 square metre building. The concept for this new plant is a combination of hi-tech auto technology, skilled human resource, and a green environment; hence, this new plant has enabled Toyota Indonesia to enhance systematic production system design in Indonesia. The production is intended for Kijang Innova, used for domestic and international markets. Besides that, this plant also produces completely built-up (CBU) units for export-oriented markets in Middle East countries (Saudi Arabia, United Arab Emirates, Kuwait, Bahrain, Qatar, Oman, Jordan, Syria and Lebanon.), while the CKD units are targeted for export to Malaysia, the Philippines and Vietnam.

The Karawang plant has been divided into three main plants. The layout of the Karawang plant is shown in Figure 7.1.

122 *Car production in Indonesia: Toyota*

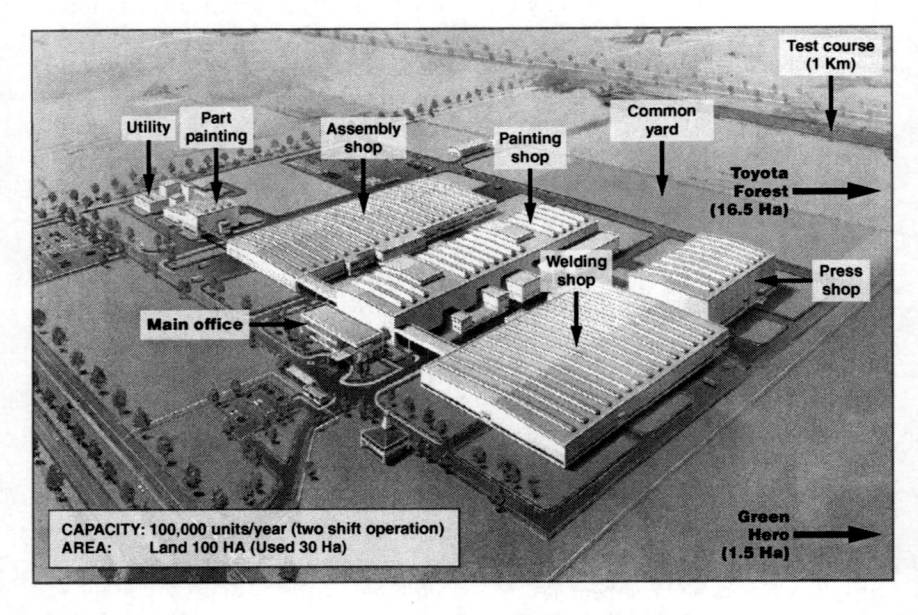

Figure 7.1 Karawang plant layout (source: author's fieldwork (2007)).

The first plant is a production plant (in this case they use the word 'shop' to have the same meaning as 'plant'), which consists of the stamping shop, welding shop, painting shop, assembling shop, test course, and common yard. Second are the supported facilities, which incluse the extended test course and learning centre. Third is the environmental facility, including waste water treatment and Toyota Forest.

The production plants in Karawang have been designed as a systematic plant based on the flow chart of the production process. This contrasts with the Sunter plants where the production layout is not a systematic model. The stamping shop, which has an area of 10,000 square metres, is where the car body is pressed. The stamping shop has two line processes which use a robotic system for higher production capacity as well as human safety.

Second is the welding shop, 23,000 square metres, where all the welding process takes place, uniting all pressed parts from the stamping shop to complete the whole car body. To guarantee the highest quality in welding, shop facilities include a welding main body line, coordinate measuring machine, and shell body line with conveyor slat. Along with 34 welding robotic system and specialist engineers, this shop is responsible for quality assurance of the completed car body.

Third is the painting shop. After the care has finished the welding process, it is sent to this shop for the electro-deeping coating and the anti-rust process, followed by body painting. Covering 17,600 square metres, this shop has 20 robotic systems to achieve high-quality body painting as well as making sure the process is environmentally friendly.

Fourth is the assembly shop, at 37,500 square metres. This carries out the finishing process to assemble the whole car component into a unit, from machine to wheels. The assembly shop has a main assembly line with doorless system assembly to guarantee the highest quality results in interior assembly and wheel alignment.

Fifth, after work is done in all five shops, the next step is the test course, which is 45,630 square metres. It tests the performance of a car from machine endurance to safety.

The final section is the common yard, a place designed for logistical purposes. Karawang has an ideal common yard that Toyota uses as a base for its delivery centre for domestic and export markets. Additionally, in support of environmental concerns, Toyota has built a waste water management system to eliminate hazardous toxins from the production process. It is 1,200 square metres, and has a capacity of 1,200 cubic metres per day; it uses flotation and anticontact aeration to eliminate sludge. Accordingly, the Karawang plant has achieved ISO 14001 for its Environment Management System.

Development of Toyota Indonesia in the automotive cluster

Japanese carmakers, including Toyota, have long had a continuous presence in the Indonesian automotive cluster, with extensive experience both in sales and distribution (Tarmidi 2004, Wee 2005, Rasiah 2005, MOTIRI 2005). This was described by the Toyota Indonesian representative as follows:

> Toyota in Indonesia will defend its position as the leading auto maker with sales growth up to 34 per cent along with the progress of Indonesian economy post-economic crisis. However, Toyota must also be cautious of market threats such as the rise of global oil which has implications for the price of raw material and several components. Despite this threat, Toyota is still optimistic to sell 500,000 units in 2008. In addition by 2009, Toyota Indonesia intends to be the base for Toyota Motor Corporation Japan in producing low-cost cars for the Asia-Pacific Rim, along with introducing a hybrid car [Toyota Prius] for the Indonesian market as a commitment to the green environment.
>
> (TMMIN representative, 23 May 2007)

As part of Toyota Indonesia's bid to become a leader in the free trade era, Toyota Indonesia will continue to export its vehicles in CBU and CKD units, engine and production tool series. Toyota Indonesia started exporting in 1987 to several Asia-Pacific countries, and even managed to export to Japan. In 2004 Toyota Indonesia exported Toyota Avanza, a collaboration product between Toyota and Daihatsu in Indonesia, to ASEAN countries in CBU and CKD forms. The export marks Toyota Indonesia's first big CBU export, which was closely

124 Car production in Indonesia: Toyota

followed by Kijang CBU units. These, as well as the export of production tools such as welding jigs and press dies since 1987, prove that Toyota Indonesia's quality meets other local customer requirements and achieves high international standards.

Toyota Indonesia's global market is detailed in Map 7.1.

Nevertheless, despite the dominance of Toyota (and other Japanese carmakers), South Korea's Hyundai and KIA are also considering setting up a production base in Indonesia to take advantage of AFTA, though there has been no confirmation yet that Indonesia has been shortlisted (MOTIRI 2005).

The carmakers (particularly Toyota) in Asia are optimistic based on current developments in the car industry (Sugiyama and Fujimoto 2000; Tarmidi 2004; Rasiah 2005). Emerging markets, particularly Indonesia, present the main opportunity for long-term car sales growth and will boost the global car market to over 60 million units by 2009 (JAMAI 2008). The prospects for car market growth in Asia are particularly positive and the Pacific Rim countries were forecast to make an additional five million units for the world market by 2009 (Fujimoto 2007; JAMAI 2008). Nevertheless, there are challenges that need to be faced. Soaring steel prices stemming from China's insatiable demand have prompted some of Indonesia's carmakers to warn of increased prices to come (Rasiah 2005; Fujimoto 2007; MOTIRI 2005).

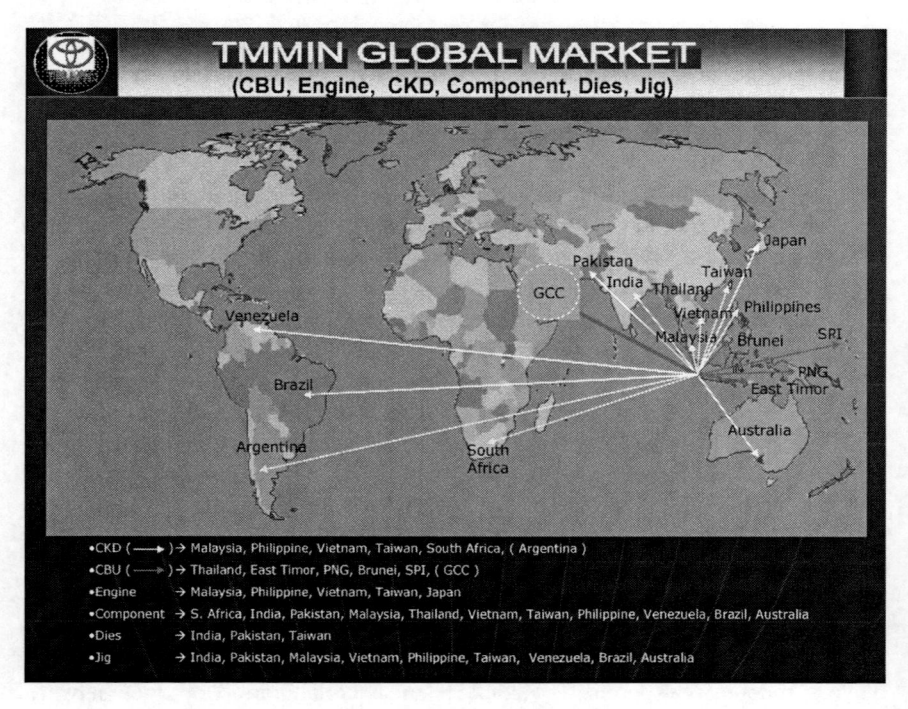

Map 7.1 TMMIN's global market, 2007 (source: modified from TMI (2007) and author's fieldwork (2007)).

Knowledge transfer within the global *keiretsu*

Knowledge transfer from Toyota Japan to Toyota Indonesia

In Southeast Asia, particularly Indonesia, knowledge transfer from Toyota has played a crucial role in developing knowledge and technology in the automotive cluster (Chen 1996; Rasiah 2005; MOTIRI 2005). Nowhere in the world is the influence of transferred Japanese technology greater than in the Southeast Asian region (Chen 1996: 7).

In addition, international technology transfer has overlaid the economic relationship between transferer (Toyota Motor Corporation Japan) and transferee (TMMIN). It might include a whole series of related issues, such as the relevant national policies and legal framework of the nation (Tarmidi 2004). Furthermore, technology transfer has played a significant role in international trade and increased the involvement of different countries in the flow of goods and services across national boundaries (Dicken 2003; Ozawa 2005; Coe *et al.* 2008).

As outlined in Chapter 3, knowledge transfer in the automotive sector by its nature is a very complicated process which involves multiple players (Hieneman *et al.* 1985; Kiba and Kodama 1991; Hatch and Yamamura 1996), because the technology normally does not have a clear-cut market value and the negotiation process is characterised by a bilateral monopoly. Therefore, the bargaining process can be intricate and difficult, with each participant trying very hard to increase its shares of the economic rents. Accordingly, trust and a cooperative spirit are indispensable for an agreement to be workable in the long run.

Technological knowledge in terms of the automotive sector in Toyota Indonesia can be classified into two broad categories: capital-embodied and labour-embodied (Odaka *et al.* 1988; Fujimoto 2007). Capital-embodied technology is intrinsic to various production processes such as casting, forging, metal-cutting, welding, pressing, etc. On the other hand, labour-embodied technology includes: (1) skills and know-how in the operation of specific processes; (2) the ability to understand capital-embodied technology, which is the ability to maintain and repair machines and equipment (this applies to elementary level employees) and (on a more advanced level) the ability to devise alternative processes and equipment in response to various economic and engineering needs; (3) the capacity to design or re-design products, processes and plants; and (4) the ability to innovate and to develop new production techniques.

Additionally, Indonesia has become not only a springboard for Toyota's products to Western European and North American markets, but is itself one of the fastest growing markets for Japanese products. Toyota's technology transfer to Indonesia has been designed to strengthen and develop its ties with this country, mentioned by the TMMIN representative in the following quote:

> Indonesia has great potential to be the base for TMC [Toyota Motor Corporation] to produce low-cost cars, regardless of other rival countries such as China, Brazil and India. However, the realisation of doing this investment has

not been officially announced yet, as the feasibility study is underway. Despite the hegemony of Tata that produced Nano along with Hyundai, Toyota wants to start the low-cost car project by 2010 or 2011. Accordingly, it is expected that the Indonesian government keeps improving its infrastructure.

(TMMIN representative, 23 May 2007)

In the automotive industry, transfer technology commonly happens between the parent company (e.g. TMC) and the host company (e.g. TMMIN). The process itself does take time as it is not simply technology per se but also involves human interaction, which in turn involves absorptive capacity (Cohen and Levinthal 1990), as mentioned by the Toyota Indonesia engineers as follows:

> When we started working in the actual plants of Toyota, it is not a trouble-free work place. The learning process would begin since the first time we joined the company and it would be a never-ending journey. The training for a new engineer will take about three months and six months for an operator under the probation scheme and they will then be examined by the supervisor and line manager to continue for a permanent contract. It is a very challenging work place and sometimes it is a stressful situation. However, once we achieved the target, it is a rewarding place to be.
>
> (TMMIN representative, 23 May 2007)

The nature of the technology that Toyota transfers to advanced industrialised countries is fundamentally different from that of the technology transferred to Indonesia as a developing country (Sugiyama and Fujimoto 2000; Tarmidi 2004; TMI 2007a). Technology transferred to the advanced industrialised countries largely consists of patented high-level technology, while that transferred to the developing countries is mainly modernisation experience and skills closely related to standardised production methods.

The scope of a typical technology transfer contract usually covers production, management, and marketing. The various production activities that TMC has transferred to Toyota Indonesia include: material selection, selection and installation of equipment, plant layout, assembly methods, machine operation, training of personnel, maintenance techniques, provision of technical data, quality and cost controls, and inventory management (TMI 2007a, 2007b). The following quote addresses the nature of technology transfer in Toyota Indonesia, mentioned by the Toyota Indonesian engineers:

> The learning process as stated in Toyota Production System [TPS] has become the fundamental foundation for TMC to transfer technology to Toyota Indonesia. However, it is not about the higher level of R&D, but it is part of continuous improvement/*kaizen* in product development as a result of market demand and customers. The final decision in R&D is a managerial level decision between the CEO/Production Manager in Indonesia and CEO/Production Manager in Japan.
>
> (TMMIN representative, 30 May 2007)

In general, one notable motive for many Japanese car producers to select direct investment as a transfer vehicle derives from the nature of the transferred technology (Sugiyama and Fujimoto 2000). For a long time, Toyota almost exclusively transferred general know-how and industrial experience (Hino 2005). The transfer of this type of technology entails long-term involvement by the transferer in the production and management activities in the host country.

Moreover, technology recipients (e.g. TMMIN) tend to require foreign investors to be involved in the initial stages of production. Many developing countries do not usually recognise the economic value of industrial expertise and tend to regard this as a free service that should accompany the purchase of machinery and equipment. Therefore, Japanese car producers found it necessary to obtain sufficient compensation for their technology through capital ownership and direct management of their foreign investment (Ozawa 1981: 40).

As most technology transferred by Japanese car producers to Indonesia is related to labour-intensive industries, labour training occupies a prominent position in the Japanese strategy of technology transfer. For this reason, on the job training (OJT) has been considered by some as Japan's 'inner mechanism of technology transfer' (TMI 2007a, 2007b).

OJT not only provides technical and administrative knowledge to employees, but also coaches them on how to have higher motivation and better discipline so that the process of never-ending quality improvement (*kaizen*) can be fulfilled. Unlike European and American companies, which utilise written manuals and detailed job descriptions, Japanese car producers support their production management methods and their technical training all through OJT (TMI 2007a, 2007b).

Additionally, in order to establish common ground for bargaining, TMC and TMMIN have to close the gaps in their price offers. This process is further complicated by some specific factors, for instance governmental regulations, political and business risk, levels of competition for technologies, and so on. Therefore, both TMC and TMMIN have to pay attention to the appropriateness of the transfer of technology as it has both macro and micro dimensions (Tarmidi 2004; TMI 2007a). The macro dimensions comprise such issues as the impact on employment and shifts in the overall balance of power among the nations involved. The micro dimensions deal with the direct impacts upon the participants of technology transfer.

Moreover, Toyota has different approaches towards technology transfer (Liker and Meier 2005, Toyota 2007). Most European and American companies will pull back their technical advisers when the factory runs smoothly, and the local employees will only need to follow manuals carefully (Miyakawa 1991; Florida *et al.* 1998; Ozawa 2005).

In contrast, in Japanese automotive affiliated companies, technical advisers tend to stay even after good operations have been achieved. They will continue to train the employees step-by-step in productivity and quality control, maintenance and repair, utilisation of new production methods and new technology, as well as other production-related skills (Chen 1996; Ozawa 2005; Fujimoto 2007).

There are a number of reasons for Toyota to adopt such an approach (Hieneman *et al.* 1985: 63). First and foremost, the technologies transmitted by OJT are basically know-how or experience related to well-proven and standardised production techniques. It has been explained by a TMMIN engineer thus:

> As technologies in the automotive sector are the type which cannot easily be transferred both in the form of industrial equipment or through blueprints or operating manuals, instead it can be better transferred through personal communication between employees and managers at all levels.
> (TMMIN representative, 30 May 2007)

Nevertheless, for most Japanese expatriates the language barrier poses a particular difficulty in communication, as most of them do not have sufficient training in local languages and their constant job rotation makes language learning even more difficult (Yamashita 1991). This problem may help explain why Japanese managers tend to like the 'learning by doing' approach in transferring technology rather than depending on comprehensive manuals that a large number of employees may have trouble understanding (Kiba and Kodama 1991; Borrus 1992; Chen 1996).

Additionally, by adopting OJT, Toyota expects to improve technology at shop floor level. As technology continually progresses to a higher level, it can hardly be written into the manual thoroughly (Fujimoto 2007). For Toyota, there should be no end to technological improvement as the company wants to be always moving forward.

Technological process is considered a dynamic and incremental process, and must be pursued by all members of the organisation rather than only by engineers (TMI 2007a). Therefore, Toyota employees on the shop floor are also involved in the activity of technological improvement (TMI 2007b). This conception is clearly manifested in the quality control that symbolises the unique strength of Japanese production management. The Japanese excel in continuing to improve the quality of their products, the process commonly called *kaizen* (Dyer and Nobeoka 2000; Liker and Meier 2005). The effect of such incremental innovation is highly visible when the product or technology is standardised (Hino 2005).

In addition, Toyota's reliance on 'transfer through people' is closely related to their emphasis on foreign direct investment (FDI) as a major channel of transfer (Fujimoto 2007; TMI 2007a, 2007b). There is usually a strong linkage between a supplier company's willingness to be involved in the training of the local employees and its financial stake in the recipient. In the case of licensing and technical cooperative arrangements, training programmes are much less significant. For example, large numbers of trainees have been sent to Japan for technical instruction under various programmes sponsored by the parent companies and the Japanese International Cooperation Agency (JICA) (Hatch and Yamamura 1996; Terry 2002; JAMAI 2008).

Nevertheless, despite evident accomplishment in technology transfer by OJT, the transfer of technology from TMC to TMMIN has not been trouble-free (TMI 2007b). The manner in which most Japanese car producers handle technology transfer has also been criticised by locals as reflecting an unwillingness to teach more sophisticated technology to the local people (Tarmidi 2004; Rasiah 2005; Wee 2005), as Japanese managers have tended to show insufficient confidence in local employees and consider it more appropriate to design and develop new products at the headquarter research centres in Japan. Therefore, Japanese car producers tend to transfer technology that is necessary mainly for routine operations. Furthermore, the heavy reliance on OJT or on the Japanese technician's experience sometimes causes serious misunderstandings between employees and managers.

Even in OJT, Toyota has encountered a serious problem, which is the relatively high rate of trainee turnover once they return to their respective companies in Indonesia (MOTIRI 2005; TMI 2007a, 2007b). Lifetime employment is not part of indigenous traditions and the commitment of employees to their companies is much less than that of the Japanese. When skilled employees return home, they are usually in high demand in the job market and find it hard to reject more lucrative offers from other companies.

Responding to this negative side, since the beginning of the 1980s Toyota has made increasing efforts to deal with the problems accompanying their technology transfer and direct investment to Indonesia (Aswicahyono et al. 1999; Tarmidi 2004; Rasiah 2005). As a result, Toyota has begun to examine the applicability of its management system and the possibility of a higher degree of localisation and decentralisation. With further diversification of production from purely labour-intensive industries to more complicated manufacturing processes, pressure has built up to expedite higher-level technology transfer. At long last, there is a mutual benefit based on understanding of the actual depiction from the shop floor to managerial decision-making. This initiative has been agreed by the Indonesian government as part of the FDI agreement and support to spur innovation in the local automotive industry (MOSR 2002; MOTIRI 2005; Rasiah 2005).

Accordingly, in the automotive industry, Japanese car producers, notably Toyota, are currently the main source of mature technology transfer to Indonesia (MOTIRI 2005). Industrial expertise and know-how have been the primary transfer while FDI constitutes the most widely used transfer channel. Traditional Japanese OJT management has commonly been used to ensure the success of the transfer process.

As a result, the process of transfer is beneficial for Toyota and its host country, since Toyota needs to shed some of its traditional industries in order to promote high-technology and service-based industries. Moreover, the continued flow of technology and investment from Japan to Indonesia will not only contribute to growth of industrialisation but also help boost the Indonesian economy in the booming Pacific Rim.

Knowledge transfer within the Indonesian automotive cluster

Knowledge transfer and Toyota's local suppliers

Toyota also wants to establish solid supply chain structures in Indonesia (TMI 2007a). As a result, forging vertical integration with the suppliers is essential. In fact, the location of Toyota Indonesia and the suppliers in the automotive cluster has been crucial to support this aim, as the JIT system must be applied in the global-local integrated production system of Toyota.

At the same time, Toyota's lean production system with its implicit localisation of supply has been diffusing more widely (Liker and Meier 2005). There has been an increase in global sourcing of mass production due to the search for low-cost high-volume suppliers within an international division of labour. This has occurred partly as a result of the customer base internationalising and seeking to source from single suppliers on a global basis, but partly from the development of specialised and proprietary technologies by certain component supplier firms among a broadening array of auto technologies (Charles and Li 1993; TMI 2007a).

Accordingly, proximity between Toyota and their business network is important in the automotive industry due to the high cost of shipping auto parts and the needs of the customers/vehicle service (Dicken 1988, 2005; Dicken and Henderson 2003). The high cost of shipping auto parts or finished automobiles also helps to reinforce the vertical integration of the auto industry. On top of that, companies are finding it tougher to compete without outsourcing secondary tasks to suppliers and focusing on their core competencies; hence, it is expected a more flexible governance system will develop (Dicken 1992; Liu and Dicken 2006).

In the Indonesian case, in terms of geographic concentration, the automotive cluster is located almost entirely in Jabotabek (greater Jakarta). In the 1990s, assembly and manufacture of the automobile was based entirely in Jabotabek, contributing 94 per cent of value added and employing 79 per cent of the workforce (Aswicahyono et al. 1999). Furthermore, the location of the components industry exhibited a similar pattern, although with slightly more spatial diversity, with 102 of the 119 major producers being located in Jabotabek, 11 in Greater Surabaya (East Java), six in other parts of West Java, and three in Central Java. Jabotabek's share of value added was 93 per cent and it represented 70 per cent of employment (Aswicahyono et al. 1999; Aswicahyono and Feridhanusetyawan 2004).

As important as proximity is, government regulations are essential to advance the industry into a globalised manufacturing set up (Moulaert and Swyngedouw 1991). As a result, through the MNEs, foreign companies are encouraged to set up manufacturing plants in the large market countries in order to reach consumers (Porter 1998b; Perdana and Friawan 2007).

A simple tiered model to describe the supply chain in Toyota Indonesia is depicted in Figure 7.2 below.

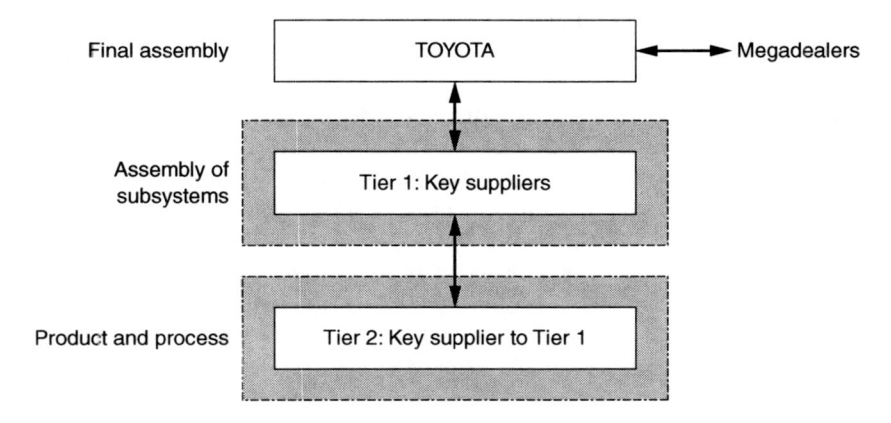

Figure 7.2 Simple model of the tiered supply chain.

Additionally, firms who produce automotive parts and components are referred to either as the manufacturers of automotive parts and components, or more simply as ancillary firms. They consist of producers of:

1 original equipment (OEM), to be installed on the newly manufactured final product
2 replacement parts or spare parts, to be used for replacing worn-out parts

Therefore, to support production in Indonesia, Toyota Indonesia must develop solid and strong supply chain networks (TMI 2007a). Thus, by developing interactive activities involving whole suppliers in Indonesia through *jishuken*,[1] Toyota Indonesia tries to maintain its first-tier supplier and second-tier suppliers, to avoid situations that might impede the continual production system in the shop floor (casting, stamping, painting, and assembling) (TMI 2007b).

In addition, vertical inter-firm linkages (the supply chain) and subcontracting networks have proliferated in the automotive industry (Sato 2001). This is to be expected. The Indonesian government's regulatory provisions have pushed firms in this direction, and there have been ample opportunities for import substitution as the local market has grown and Indonesia's industrial competence has developed. As a result, the range of locally produced components has risen dramatically since 1970 (Aswicahyono *et al.* 1999; Sato 2001; Tarmidi 2004).

By the mid-1990s, local production had extended to many items: filters, pistons, petrol and diesel engines, axles, transmissions, steering systems, safety-glass, high-quality forged parts, wiring harnesses and aluminium wheel rims, in contrast to the 1970s, when very few components were produced (Sato 2001; Rasiah 2005). By the early 1980s the process of backward integration was beginning, but it was delayed by the mid-1980s recession, when the government also relaxed its local content requirements for several years (Rasiah 2005).

In general for the Indonesian context, the nature of the car producer–supplier relationship has varied greatly. Much of the official thinking in the 1970s envisaged Japanese-style subcontracting networks, characterised by stable, durable, and intense ties between the two groups of firms, in which the car producers would be the dynamic agents of technology transfer to clusters of small-medium components manufacturers (Sugiyama and Fujimoto 2000; Tarmidi 2004).

Nevertheless, in contrast to other Japanese carmakers (Suzuki, Mitsubishi, and Isuzu), the more common pattern is for both parties to engage in opportunistic behaviour whereby relationships are typically shallow, short term and non-exclusive. The reasons for these differences are not difficult to explain (Hino 2005; Fujimoto 2007). With a highly segmented and small market, supplier firms for non-Toyota groups in Indonesia are seeking to achieve scale economies and thus are forced to serve several carmaker firms (TMI 2007a, 2007b). Accordingly, these carmakers have tended to eschew long-term relationships with myriad domestic small-scale producers of low-quality replacement parts. As noted, foreign investors who are unable to secure majority ownership are reluctant to transfer technology along with weak and sporadic government industrial extension efforts.

Additionally, non-Toyota companies in Indonesia possess limited technological capacity and therefore are not in a position to act as agents of technology transfer. Indeed, in one particular case the technology flows operate in the reverse direction, especially where the suppliers are in joint ventures with foreign companies. However, some argue that Indonesia's business culture with its emphasis on short-term, opportunistic behaviour is not well suited to the establishment of longer-term ties (Tarmidi 2004; Rasiah 2005). How deep-rooted this characteristic is, and how much it reflects specific features of the business environment during the Suharto period, is a subject for future research.

Subsequently, the government's policies could hardly be judged successful (Sugiyama and Fujimoto 2000; MOSR 2002; MOTIRI 2005). Subcontracting relationships will flourish, not by bureaucratic mandate, but rather through market-driven incentives (Rasiah 2005). The policy environment (i.e. fragmented markets), high protection, and weak industrial extension efforts have in fact been inimical to the development of such ties. It is therefore not surprising that forecasts of spillovers from foreign to local firms are apparently quite limited (Tarmidi 2004; Aswicahyono and Feridhanusetyawan 2004). However, the Toyota group is a notable exception.

In fact, auto industry suppliers in Indonesia consistently report that Toyota is their best customer and their toughest. 'Toughest' here defines Toyota as having high standards of excellence, and the company expects all their partners to rise to those standards, as expressed by the ATI (Aishin Takaoka Indonesia) engineers in the following quotes:

> There is no doubt that Toyota has dominated the supply chain from mostly local auto component makers as well as international supply. Under Toyota Group, it is possible to have business networks with any vendors. In turn,

having business deals with Toyota is also a business opportunity for the supplier to strengthen the position as solid auto player. Every company is competing to supply for Toyota either by being first tier or second tier.

(ATI representative, 12 May 2007)

In Indonesia, Toyota has been recognised and respected among its suppliers for serious investment in building a network of highly capable suppliers that are truly integrated into Toyota's extended lean enterprise. Toyota suppliers are integral to the JIT philosophy, both when it is working smoothly and when there is a breakdown in the system.

Toyota Indonesia has followed the principle from its parent company in Japan regarding the need to find solid partners and grow together in achieving mutual benefit for long-term cooperation in both enterprises (TMI 2007b). Working together towards common goals is the heart of what has made Toyota's partnerships with suppliers a global benchmark.

The familiar cost-cutting reason is not the only one. Toyota Japan wants to invest in Asia, in particular Indonesia, as a potential market as well as long-term supply chain management. Additionally, Toyota takes into consideration other reasons for preferring local sourcing: the operating philosophy of JIT, pressure from the Indonesian government, a philosophy of positively contributing to the social and economic development of Indonesia, and to encourage growth and prosperity among employees, dealers, and suppliers. Accordingly, Toyota Indonesia perceives this matter as a complex issue with multiple ways to respond (Liker and Meier 2005; personal interviews, Toyota, Toyota's supplier, 2007).

First, Toyota Indonesia sees quality as more than just having state-of-the art equipment and ISO 9000 documented quality procedures. It begins with the people doing the value-added work. In short, Toyota wants its suppliers to have a compatible culture of finding and eliminating problems through continuous improvement. Treating suppliers as partners in the business way can aid long-term success.

Second, there is close and integrated engineering of products and processes. Toyota has made a living from the overall quality of design, precision, and flexibility of its manufacturing processes. The integration of product and process in the design and engineering stages has a huge impact over the life of a product. Since suppliers are manufacturing the components, high-quality design and manufacturing can be best done in close concert with, or even by, the suppliers. Thus, integrating engineering between Toyota and its supplier in product and process is the critical success factor and takes many years of investment to get it right. In Indonesia, this process started in the 1970s when Toyota Indonesia established its first assembly plant in Jakarta.

Third, there is precision and delicacy in the JIT system. Having suppliers who do not have this capability creates weak links throughout the value chain. Thus, Toyota wants every link to be equally strong and capable in order to make connected flows between stable processes.

Fourth, Toyota wants innovation. The core of its long-term success has been through innovation – in products, processes, and countless small improvements throughout enterprises. Toyota sets specific targets for innovation by its suppliers. In addition, working with suppliers to set specific plans for more investment in R&D to put innovative technologies on the shelf means the technologies are proven and ready for chief engineers to pick off the shelf to design into mass-produced vehicles.

Fifth, Toyota realises that the overall financial health of the Toyota Enterprise depends on the overall financial health of each part of the enterprise. Therefore, a weak supplier may be able to inspect and build inventory and ship good parts JIT and provide price reductions, but at some point the weak supplier will be driven out of business. Accordingly, Toyota wants suppliers that are strong and capable of contributing to the entire enterprise.

In general, 103 listed suppliers of Toyota Indonesia said that they prefer working with Toyota (and similar customer Honda) to any other car company due to the well-maintained supplier partnership in the Toyota Way (personal interview, Toyota 2007). However, for suppliers to collaborate with Toyota is not an easy task. Being a partner to Toyota is not about Toyota being soft or forgiving. As pointed out in the Toyota Way, fairness, high expectations, and challenge characterise how Toyota treats suppliers. The phrase 'long-term relationship' connotes trust, caring, and mutual well-being, yet also signifies discipline, being challenged, and improvement.

The domination of the Toyota business network in Indonesia spreads across the 33 provinces, from Sumatra Island to West Papua. However, the strongest ties are located in Java as the centre of Toyota in Indonesia. Under this network, Toyota has been consistently using its *keiretsu* in the name of the Toyota group to strengthen its position as the leading car manufacturer in Indonesia. Even if another carmaker wanted to take over Toyota's position, it seems unlikely they would be able to, as Toyota has embraced and approached nearly half of the local and international auto components supply for its supply chain in Java.

Toyota Indonesia has adopted from Toyota in Japan the seven characteristics of supplier partnering as follows:

1 **Mutual understanding**
The basis for the relationship starts with understanding and it does not come easy. For Toyota, *genchi genbutsu* (actual part, actual place) is the core philosophy of going and seeing directly, to deeply understand a situation.
(TMMIN representative, 23 May 2007)

2 **Interlocking structures**
In here, technical systems, social systems, and cultural systems are all tightly intertwined. The supplier must act as a seamless extension of the refined lean system of Toyota. The structure reinforces the interdependent processes with Toyota.
(PT Taiho Nusantara Indonesia, 20 April 2007)

3   **Control systems**
    Toyota is command central for supplied parts. They know the status in real time of all parts suppliers. Toyota has a database consisting of any key delivery performance indicator for any supplier, performance over time on quality, cost, and production control. If there is a near missed shipment, a quality problem, incorrect labelling, or any glitch, it will surface immediately. Control also extends to aggressive cost reduction initiatives.
    (TMMIN representative, 30 May 2007).

Along with that comment, another view:

    Toyota not only gives the supplier a target, but also carefully monitors progress in reducing costs to achieve targets. Toyota has parallel sourcing which means source not from one but not from many. Toyota looks for three of four top-notch suppliers for a component and keeps the business within this family [i.e. Toyota Group]. For any given car model, one of the suppliers will get this business for the life of that model. Nevertheless, getting it for the next version of that model is not guaranteed. If they do not perform, or their competitor, like a sibling, does a lot better, they can lose this business.
    (PT Nippon Piston Ring Indonesia representative, 25 May 2007 and PT Asama Indonesia representative, 22 May 2007)

4   **Compatible capabilities**
    In low-wage countries like those in Asia (Indonesia, China, India, and Thailand), Toyota views suppliers as an extension of its technical capabilities. It is Toyota's development process. Thus, Toyota continues to invest heavily in teaching Indonesian engineers; hence, their way and the capabilities are gradually building in Indonesia. At the moment, Toyota has stepped out its simultaneous engineering initiative, getting input from suppliers on their manufacturing when it is still a concept and before the body is even styled.
    (TMMIN representative, 30 May 2007)

5   **Information sharing**
    Toyota believes strongly in information sharing. There is a high degree of structure with a specific time and place for meetings, very clear agendas, and clear formats for information and data sharing. Designing requires the most intensive level of supplier involvement and must be done on Toyota's CAD system.
    (TMMIN representative, 30 May 2007)

The idea is that suppliers design their components into Toyota's vehicle. However, separable body functional parts are designed fundamentally by the suppliers in their own buildings. They are referred to as RDDP [Request for Design and Development Process] parts and could be done on the supplier's system with less intense communication.
    (PT ATI representative, 23 May 2007)

6   **Joint improvement activities**
    Most of Toyota's suppliers perceive the contract with Toyota as opportunities to learn and get better and enhance their reputation with other customers. Toyota does not just purchase parts from suppliers, but also develops supplier's capabilities. Toyota's goal in teaching its suppliers lean methods is not to teach specific tools or methodologies, but to teach a way of thinking about approaching problems and about improving processes (i.e. the Toyota Way).
    (PT OTICS Indonesia representative, 12 May 2007)

7   **Continuous improvement and learning**
    Through *jishuken*, Toyota and its suppliers share a common philosophy of product development and manufacturing, and many specific practices.
    (PT ATI representative, 12 May 2007)

In the co-development product, it is essential to be completely synchronised on timing, testing methods, metrics to specify product performance, and even technical vocabulary. The result has been the evolution of common philosophies, language, and approaches between Toyota and its suppliers.
(PT Menara Terus Makmur representative, 20 May 2007)

In addition, Toyota Indonesia has honed its skills in applying TPS by working on projects with suppliers. Toyota Indonesia needs its suppliers to be as capable as its own plants at building and delivering high-quality components JIT. Moreover,

Toyota Indonesia cannot cut costs unless suppliers cut cost, lest Toyota simply push cost reductions onto suppliers, which is not the Toyota Way. Following the parent company in Japan, Toyota Indonesia does not view parts as commodities to be sourced on the market through open bidding. Therefore, it is critical that it works with highly capable suppliers that are following TPS or an equivalent system.
(TMMIN representative, 8 June 2007)

As a result, Toyota Indonesia uses interactive methods to learn with its suppliers and in the Toyota Way style. These are all learning by doing processes, keeping classroom training to a minimum. The important learning happens through real projects on the shop floor.

All key suppliers are part of Toyota's supplier association. These are core Toyota suppliers that meet throughout the year in the Head Office in Jakarta or by visiting the supplier's plant to discuss the appropriate problem-solving based on no good part or complaint. Members of Toyota's supplier association can participate in many activities, including study groups that meet to develop greater skills in TPS. These are called *jishuken* or voluntary study groups.
(TMMIN representative, 8 June 2007)

The Operations Management Consulting Division (OMCD) started this *jishuken* in 1977 in Japan. OMCD is the elite corps of TPS experts started by Ohno in 1968 to improve operations in Toyota and its suppliers. In Indonesia, *jishuken* was introduced in the 1990s when Toyota insisted on improvements to its suppliers' quality, cost, and delivery systems. The *jishuken* identifies a business need and then picks a product line to do a project. This project consists of developing a 'model line'. A typical model line includes component assembly and a manufacturing process that makes parts that go to the assembly line. A full TPS implementation is done with all the elements of JIT, *jidoka*, standardised work, Total Productive Maintenance, etc.

Nevertheless, *jishuken* by design is not part of the business relationship with suppliers. It is there to educate through project work. Toyota purchasing has its own quality and TPS experts work with suppliers when there are problems. The most severe problem would be if a supplier shuts down the Toyota assembly plant because of a quality or production problem. Thus, Toyota must save its 'sick supplier' through TPS. If a supplier puts a Toyota assembly plan in danger of shutting down, there will be two emergency solutions. Level one, Toyota will then send a team of people swarming through the supplier's plant and the supplier must develop an action plan to address all of their concerns. Level two typically means severe probation for a year. As a result, to supplement the existing *kanban* lean production and delivery, Toyota Indonesia has been using *e-kanban* since late 2005 to its suppliers along with a milk-round system, particularly for those suppliers located far from Toyota's plants in Jakarta or Karawang (TMI 2007).

Furthermore, Toyota Indonesia has been developing its supplier databases to maintain progress and find out quickly if a new supplier is needed.

> Toyota Indonesia looks at its suppliers based on part flow. The first tier is the supplier who has a close relationship with Toyota in making certain parts. This first tier is also in charge of maintaining the involved second tiers who produced the components of those parts.
> (TMMIN representative, 30 May 2007)

> As in automotive industry, one component is made of lots of parts. Therefore, to join them together, Toyota gives the responsibility to its first tier. This part flow is so complex and contains many and different suppliers locally and globally (Toyota Indonesia mainly receives overseas suppliers from Asian countries). Thus, it is also possible that first tier can be second tier, and vice versa.
> (TMMIN representative, 8 June 2007)

Accordingly, communications among the suppliers are well connected, principally for those suppliers under the Toyota Group. There is cooperation among the suppliers to learn from each other in order to maintain the common goal, supplying parts to Toyota.

The need for close proximity between component supplier and assembler is required in order to reduce inventories and waste as well as improving quality control. Accordingly, component production would be likely to become more tightly concentrated at the point of final assembly in the major markets.

(TMMIN representative, 8 June 2007)

The type of geographical concentration of component suppliers that occurs around Toyota's two plants in Java would be the ideal model for an endeavour to 'duplicate' the parent company in Toyota Japan (TMI 2007).

Furthermore, as the vehicle system incorporates new technology in both product design and productive processes, these would be increasingly manufactured near to the supplier's centralised production location. As a result, closer involvement between workforce and shop floor management would give potential for higher quality in production and a source of new ideas for product/component design and the productive process itself. Low inventories would mean less inactive capital, while a more involved cooperative workforce should indicate less middle management. The spatial dispersion of TMMIN's supply is described in Map 7.2.

Kijang Project: Toyota Indonesia's localisation project

It is increasingly apparent that car producers in newly-industrialising countries cannot just rely on traditional, labour-intensive manufacturing operations, particularly if they want to enhance their bargaining power by being prominent in automotive export markets.

According to Jonash and Womack (1985: 405), if newly-industrialising countries are intent on developing an export-oriented motor vehicle industry, the objective should be to maximise the amount of wealth produced per worker using state-of-the art social organisation and production technology, and to create jobs by expanding volume.

If this does not take place, Jonash and Womack (1985) argue that newly-industrialising countries will fall further behind in technological development and have even less chance of ever catching up. As a result, the implication of such analysis must be that Indonesia, regarded as a newly-industrialising country, has to forge stronger links with Japanese vehicle multinationals, particularly in terms of technology transfer and new organisational methods.

Toyota Indonesia wants to take up this challenge not only through assembly line plants but also by active involvement in producing engines, parts and components for export-oriented production specifically in Asia-Pacific, the Middle East, Africa, and Latin America. One of the preferred types of passenger car for exports is the Kijang.

Kijang (deer in English) is one of Toyota Indonesia's accomplishments through an extended pathway. A Kijang prototype idea was designed by Toyota Indonesia's engineers based on push factors: the Indonesian market, local

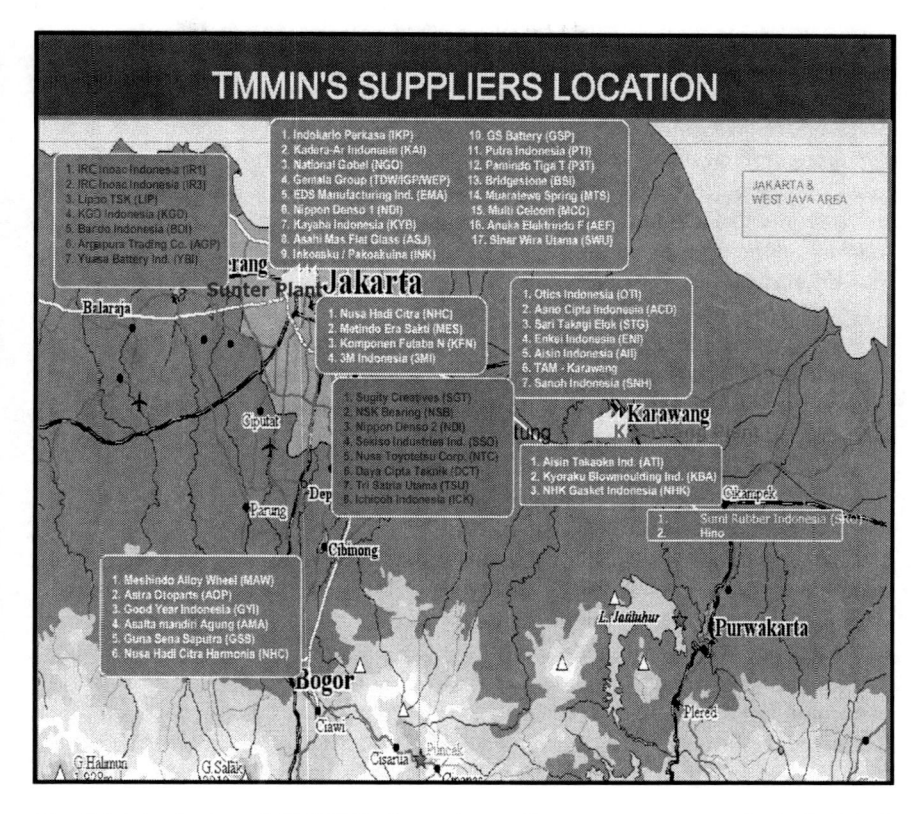

Map 7.2 Distribution of TMMIN's domestic suppliers, Java (source: author's fieldwork (2007)).

context, the urge from government to use 80 per cent local components. However, as Toyota Indonesia was not the final decision-maker on whether to proceed with this project, the initial design was sent to TMC Japan for further feasibility study and improvement. This process involved Indonesian engineers in the Kijang team project to come up with more details for production and technology improvement alongside Japanese engineers in Japan. After a long ongoing improvement process and rapid strategic marketing to target the 4x2 segment, this Indonesian family car is a phenomenon in the history of the Indonesian automotive industry. Kijang's reputation has soared from its debut up to now, with a place in the heart of every Indonesian family. Furthermore, Kijang production reached the one million mark by the end of 2003; hence, Kijang was recorded in MURI (Museum of Indonesian Records), which strengthens Kijang's reputation as the best Indonesian family car.

The first generation of Kijang began in the period 1977–1980. Kijang's debut on 9 June 1977 was extraordinarily well received. The second generation of

Kijang was launched in 1981–1986. Since then, Kijang has received an even more favourable response. In fact, the growth of the coachwork business soon afterward was a key element in the transformation of the Kijang from a commercial vehicle to a family car.

The third generation of Kijang was sold in 1986–1996. Furthermore, Kijang's popularity increased when Super Kijang was introduced in 1986 and Kijang Grand in 1996 with the Toyota Original Body technology. As a result, Kijang started to be caulk free and has the same body structure quality as the sedan. It has made Kijang overtake the sedan as the most popular family car. Therefore, in 1995, production of Kijang reached 500,000 units and the introduction of a 1,800 cc engine was finalised.

In the fourth generation, the new look Kijang was introduced as Kijang Kapsul, with 60 per cent local content. This new generation of Kijang was very different in design from the previous one. It also offered a number of variants, including the choice of gasoline or diesel engine, the luxury Krista and the option of automatic transmission.

Early in 2000, Kijang Electronic Fuel Injection was introduced, followed by a more powerful 2,000 cc EFI Engine. In the fifth generation, Toyota Indonesia proudly introduced Kijang Innova, a revolutionary design that continues the ultimate performance of Kijang. With its stylish revolutionary look and new improvements in all features, Kijang Innova offers a significant driving experience.

Accordingly, in response to global expansion, Toyota Indonesia has been exporting Kijang since its launch. However, the most significant exports of Kijang began in 1998, particularly to ASEAN and Middle East markets.

Links between Toyota Indonesia and the university

As part of its existence in Indonesia and basic business philosophy that Toyota builds people,[2] Toyota Indonesia demonstrates this respect by providing welfare for employees, business partners (particularly in Toyota's *keiretsu*), and stakeholders, and by contributing environmental, social-economic, and technological benefits to the host country. Those efforts have also been considered part of Toyota's strategic plan in terms of a sustainable long-term business agreement in Indonesia.

With regard to knowledge transfer, with an emphasis particularly on automotive engineering, Toyota Indonesia along with Astra Indonesia, has undertaken several educational projects. This is a usual channel for the MNEs through FDI in Indonesia to contribute 'something' for the host country, as requested by government regulation and business contracts, but also as a part of the company's business philosophy.

Astra International Indonesia is one of the most prominent business conglomerates in Indonesia. Their businesses consist of automotive, financial services, heavy equipment, agribusiness, information and technology, and infrastructure. Astra's automotive business comprises the production, distribution, sales and aftersales service of automobiles, motorcycles, and automotive components. Through partnerships with Toyota, Daihatsu, Isuzu, Nissan Diesel, Peugeot and BMW in the automobile sector as well as with Honda in the motorcycle sector, Astra is Indonesia's leading firm in marketing and distribution in the automotive sector.

Together with Astra International, Toyota Indonesia has established the Astra and Toyota Foundation (ATF), whose mission is to participate in the education sector in Indonesia by offering scholarships from elementary school up to Master's degree level, as well as providing funding for scientific research, teaching aids, and training.

In 2006, ATF granted scholarships for over 5,580 elementary, junior high and senior high students, 950 university students, and 175 polytechnic students, as well as 16 postgraduate students. Aside from scholarships, ATF also donated support for student scientific activities in 21 universities and for 14 postgraduate students who were finishing their final research theses.

Moreover, ATF donated various engine, transmission, and other automotive components as well as automotive books, and other learning tools to support technician training programmes in various educational institutions across Indonesia.

The current activities have been taking place in a school–university–industry link scheme across Indonesia, called Toyota Technical Education Programme (T-TEP). It is based on the consideration that job experience for students has become more important in a world of education, particularly for those who have a background in vocational schools in industry. Therefore, through one of the various programmes run by Toyota and Astra, such a necessity will be fulfilled.

T-TEP is a programme that functions between these two main areas of education and industry, and has been running since 1994. T-TEP undertakes various activities, such as training for students, training for teachers, the donation of school outfits to vocational students in Jakarta and its surrounding area, as well as donations to international scope schools.

In addition, on 22 June 2007, T-TEP held an opening ceremony of the Double System Programme (DSP), which aims to prepare employees from industrial backgrounds. DSP as a national scope programme already participates in several schools in Jakarta others from all over Indonesia.

Accordingly, before joining DSP, students have to pass the selection steps, starting from their own school. Each school has a quota of 20 students, who are considered to be the best students in its school. Then the next step is for those 20 students to compete with other selected students from all school participants, for the opportunity to join the training scheme. The students will join this training for one year or whatever is necessary for the Toyota dealers in Jakarta and its surrounding area (Jabotabek).

At the end of training, a student who passes the competencies test will get an official certificate from Toyota Indonesia. This certificate gives them priority to join Toyota's dealers after they graduate from school.

After being successful with the T-TEP programme, ATF launched a continuation of the T-TEP programme called Sub T-TEP. Both programmes intend to improve potential human resources by familiarising students with the latest automotive technology. Both programmes receive the same facilities such as training opportunities for teachers, training manual books, CD training, school curriculum adjustments in line with the Toyota curriculum, and on the job training (OJT) opportunities in a Toyota workshop. As a result, the students who graduate from Sub T-TEP will be of the same skill-standard as those who graduated from T-TEP schools. Furthermore, Sub T-TEP could help technology transfers reach a wider area. A year after the Sub T-TEP launch in 2006, Toyota Indonesia officially linked up with another 13 vocational schools as Sub T-TEP institutions.

In addition, T-TEP and Sub T-TEP have also become centres of skill education development for surrounding vocational schools. This provides development through the latest technology information transfer programme, training, and improvements to environmentally friendly technology, as Toyota wants students to build their environmental awareness before entering the industrial world.

In the latest news, Sub T-TEP institutions will be increased, with another ten vocational schools, to be followed by another 110 vocational schools. Moreover, Sub T-TEP will be developed in a vocational schools development pilot project in each province. Then it will become the centre of activities for technology transfer, Toyota new models and new mechanisms of knowledge transfer, and as the place to hold the competency test for students and teachers from surrounding Sub T-TEP programmes. As a result, Sub T-TEP finally would strengthen its position in a cooperation programme between ATF and the National Education Department.

Along with vocational schools, ATF has also targeted academia by visiting universities across Indonesia. ATF university visits have been regarded as important to academics for making better links with industry. It is also expected that ATF will be elaborating its entrepreneurship training and student exchange programme with another of Toyota's plants in ASEAN countries.

Additionally, a similar programme has also been undertaken by Astra with help from Toyota Indonesia in providing technical training assistance to SMEs, particularly those that serve Toyota's group, both vendors and subcontractors. To roll out the programme, local governments, state-owned enterprises, private companies, and universities have cooperated to develop nine Business Development Centres that assist 1,560 SMEs in several provinces, which consist of Jabotabek area, West Java, Tegal, Jogjakarta, East Java, Bali, Nusa Tenggara Barat, Borneo and Sumatra.

Along with that programme, capital funding help is also provided. Since 1991, a total of Rp101 billion has been invested in SMEs to expand their business from automotive components, motorcycle spare parts, heavy machinery, rubber and plastic manufacturing to handicrafts and production houses, as well as paint and medical equipment manufacturers.

Subsequently, Astra has another educational programme with support from Toyota. It is the Astra Manufacturing Polytechnic (Polman Astra) and BERNAS

(Educational Assistance and Transformation for Children and Schools). Polman Astra aims to develop effective learning concepts and to ensure that its graduates gain employment. Polman Astra has developed close links with industry, particularly with the Astra Group and also with the Group's customers and suppliers. For example, Polman Astra collaborated with United Tractors to launch its new Heavy Equipment Programme and also with Astra Nissan Diesel Indonesia by sending some of its staff and alumni to Nissan Diesel, Japan. The school works closely with various national and regional government and non-government organisations, including the Ministry of Education, the Japan Vocational Ability Development Association (JAVADA) and the Association for Overseas Technical Scholarship (AOTS), which has appointed Polman Astra as its Indonesian Centre of Excellence in Manufacturing.

In its curriculum, Polman Astra has five core programmes. They are manufacturing machining technology, manufacturing process and production technology, information system for manufacturing, automotive technology, and combined engineering technology (including electronic, information, mechanical, and control system engineering). All these programmes are supplied with a supportive laboratory and the latest machine and engineering software.

Polman Astra's ongoing role in R&D has yielded various successes, including the development of a Wireless Application Audience Response, Plasma Cutting machines and retrofitting for CNC machines. The school is proud to announce that one of its students on the Production Technique and Manufacturing Process Programme was selected to represent Indonesia in the Asian Skill Competition VI – Computer Aided Design & Drafting (CADD) in Brunei Darussalam.[3]

Furthermore, BERNAS assists schools in less-developed regions through human resources, curriculum and school management development and donations for teaching aids, as well as initiatives on life values and life skills development. In 2006 BERNAS completed the renovation of an elementary school and provided assistance to four other elementary schools in Leuwiliang, West Java. To improve the competence of the headteachers and teachers, BERNAS also organised a training programme on teaching and curriculum development skills and donated books for the students, teachers and libraries, as well as various teaching aids.

Conclusion

The globalisation of Toyota in the Indonesian automotive cluster has caused restructuring, developing it into a truly worldwide organisation. This initiative has taken place in the Jabodetabek area across the Java region since the 1970s. Since then, the operation of automotive global production knowledge has been progressively growing in the Java automotive cluster, along with managerial and technological path dependency. This process has been taking place as part of the knowledge transfer activity which involved the suppliers in the automotive cluster across Java.

144 *Car production in Indonesia: Toyota*

The decision to locate in the two clusters in Java has made Toyota Indonesia become the pre-eminent example of how an Japanese automotive MNE can spill over its knowledge and technology to the local economy. This should be added to supplement the currently insufficient literature on knowledge transfer in automotive clusters from developing countries.

Being located in the cluster has provided benefits for Toyota, both in terms of external linkages (international strategic alliances and the development of new supply chain structures) and internal linkages (team working, network management structures, flexibility, etc.), as well as accelerating the growth of the automotive cluster in the Java region. The benefits are:

1   Promoting an investment climate conducive to business by strengthening cluster growth within the automotive industry in the Java region
2   Providing assistance to the automotive industry to penetrate the global market as well as to upgrade competitiveness by accelerating the industry
3   Enhancing competencies of human resources for the automotive sector
4   Liberalising the auto and auto parts market
5   Accelerating the use of e-commerce among suppliers, as giant automotive manufacturers establish databases for supply chain management

The establishment of vehicle and engine plants in Indonesia by Toyota has led to a net increase in demand for locally supplied components. It is also part of Japanese production systems (i.e. TPS) which use the concept of lean production by JIT and *kanban* systems, to avoid waste in manufacturing activities. In addition, these plants are likely to be organised on quite stringent JIT principles, thus reinforcing the need for new component supply sites close to vehicle assembly plants.

Toyota Indonesia and Astra International Indonesia have been working together to develop the knowledge transfer programme in Indonesia. OJT is the appropriate knowledge transfer channel for the Indonesian context, which is more focused on human interaction and practical ability. Accordingly, various educational activities run by Toyota Indonesia have provided social-economic benefits for the host country.

8 Motorcycle production in Indonesia

The Honda complex

When Honda invested abroad, its overseas activities became subject to a great deal of interest and speculation. Could Honda really function in Indonesia as well as in Japan? Would this mean the spread of Japanese business culture? As Honda argues for the significance of developing technology in a global perspective, what would that mean for the Indonesian automotive cluster as the host country?

In the Honda case study, the distinctive finding is the process of knowledge transfer between Honda and its suppliers through New Honda (NH) circle activity and the importance of being located in the same automotive cluster. This chapter provides a case study of Honda Indonesia to address the following research questions:

1 **Outcomes:** How has Honda contributed to the development of the Indonesian automotive cluster?
2 **Process:** How has Honda been fostering technological change and knowledge-technology transfer in the Indonesian automotive cluster? Particularly with reference to the role of knowledge transfer in the Honda supply chain, and considering the role played by Honda as a leading multinational enterprise (MNE) to drive the progress of the automotive cluster by locating its transplants there.

Chapter 8 begins with a description of the internal workings of Honda and the role of the Honda subsidiary in Indonesia. It will then explain the Honda supply chain (vertical network) and the process of knowledge transfer through the role of suppliers in the automotive cluster in Java region. Additionally, the case study also describes the Revo Project as a localisation project from Honda Indonesia and the relationship between Honda Indonesia and the local academic institution.

In taking an in-depth look at Honda's multinational operations, the analysis is forced to overturn many commonly shared assumptions about the Japanese firms. By looking at the Honda case study in Indonesia, the analysis reveals that Japanese business has its own unique ways in its overseas operations, in particular the capacity of Honda to innovate and transform its overseas operations and be responsive to the host country. Accordingly, it is necessary to look closely at Honda's global operations to discover how to really make sense of the Japanese MNE and to dispense with many of the enigmas involved.

The Indonesian subsidiary within the Japanese automotive MNE

Honda's internationalisation strategy

Honda had both two-wheel and four-wheel divisions and now has a third division of power products that stem from the entrepreneurial spirit of its founder (Mito 1990; Mair 1994; Honda 2004). In support of the motorcycle division, Honda has several reasons why this remains an important division. Motorcycles have become the main alternative mode of transport, as an easy-rider for going out with friends, getting a direct feel for the rhythms of the city, as well as running errands or doing business. That is why Honda make scooters, electric motor-assisted bicycles, sports bikes, large touring cycles and more.

In 1949, Honda manufactured the first commercial motorcycle, the 'Dream Type D' in Japan. In 1963, Honda opened its first overseas plant, in Belgium. Since then, Honda has followed one basic rule: build products close to the customer (Sakiya 1982; Mair 1994). The result has been a worldwide succession of manufacturing facilities for a total of 28 motorcycle plants in 21 countries, as well as Honda R&D operations in the USA, Germany, Italy, Thailand, China, and India – all working to develop motorcycles that match national needs.

Accordingly, along with its global strategy, Honda's safety net is to have a complementary policy in product development for both motorcycle and car divisions (Honda 2004). Motorcycles and cars are both vehicles with engines but they are quite different machines. Between them, the two-wheeled and four-wheeled vehicles can provide a strong safety net in the automotive industry.

As a central theme in Honda's global strategy, Honda Motor Corporation (HMC) moved towards a more global system of management. This can indicate that any top-level problem in Honda is bound to be complex (Honda 1991, Mair 1994, Honda 2004) as a Honda Indonesia representative confirms:

> Particularly in establishing or remodelling a new plant, decisions must be made about the size of the plant and its location; what products are to be manufactured there; how many workers will be required; what will be the cost of the investment. All these matters must be examined carefully.
> (Astra Honder Motor, AHM, representative, 27 April 2007)

For that reason, Honda intends to go beyond the 'localisation of production', which means building plants in other countries and employing local workers, and to do the 'localisation of management' (Honda 1991; Mair 1994; Honda 2004).

> What is even more important is to have an accurate information and understanding of world market trends (to compete with major rivals). Therefore, all these decisions are made in so-called joint board rooms between management of Honda Motor Corporation and Honda Indonesia.
>
> (AHM representative, 29 May 2007)

Having a mission to be the global-local corporation, Honda heralds an industrial revolution in global-scale manufacturing with profound economic, political, technological, and social consequences for the countries where it operates (Honda 1991, 2004; Mair 1994). It was a challenge for Honda to operate outside Japan because of the assumption that it would not be able to repeat its home-based success.

As external economic conditions have become more complex, the complementary nature of motorcycle and car production and sales has proved to be precious for Honda. Another example of Honda's complementary policy is found in the relationship between its domestic and overseas markets. The Japanese motorcycle market had reached maturity by the mid-1960s, which meant that the growth of the motorcycle industry heavily relied on the overseas market (Honda 1991, 2007a), particularly in an emerging region such as Southeast Asia.

> Studying the characteristics of each country and its market enables one to identify the significant differences between them as well as ways in which they might complement each other. Each market has its own historical and cultural features; they are not uniformly influenced by macroeconomic trends nor do they respond to them in the same way. Within Southeast Asia, and specifically Indonesia, there are countries with different political and economic systems as well as distinctive cultures. Also, they differ in terms of economic and social development.
>
> (AHM representative, 29 May 2007

In the Asian region generally, demand for motorcycles as a principal means of transportation has continued to grow year on year since 2000 (Kiba and Kodama 1991; Gunawan 2002). Total unit sales of motorcycles and motorcycle parts sold by Honda and its subsidiaries to affiliates in the region jumped 18 per cent to 7,017,000 units in 2004 (Honda 2004).

In Southeast Asia, Honda's largest motorcycle market, the economy bottomed out in the middle of 1999 (Honda 2004, 2007a). While there has been an influx of counterfeit models into this region, Honda has effectively maintained favourable sales in such large markets as Indonesia, Thailand, and Vietnam with a variety of family-sport type models (Honda 2004; JAMAI 2008).

Seeing that motorcycles remain the most prevalent mode of daily transportation in this region, Honda must continue to introduce new models quickly and with improved cost competitiveness. In view of that, Honda is working hard to expand local businesses in the region through its active promotion of local parts procurement used in overseas production, so-called 'localisation' and being situated in the local automotive cluster (Mito 1990; Mair 1994; Honda 2004), as confirmed by Honda Indonesia's representative as follows:

> The external economies policy of each country is often different so that market accessibility is not the same. Therefore, it would be unwise to regard the ASEAN and Indonesia as one market. In fact, each country calls for close analysis, marketing tailored with great care. In sum, Honda Indonesia wants to be the global player in the region and use the local taste to enter the market.
>
> (AHM representative, 29 May 2007)

This strategy has resulted in a sharp increase to 100 per cent locally procured models made by Honda affiliates in India and China that are not included in Honda's unit sales, followed by Indonesia, which is still in the region of 90 per cent for locally procured models (Honda 2004, 2007a). Nevertheless, production and unit sales of 100 per cent locally procured Honda-brand motorcycles in Asia rose significantly from the preceding term, to around 1 million units (Gunawan 2002; Honda 2007a).

Establishment of Honda Indonesia in the automotive cluster

The Honda motorcycle in Indonesia was first established in the Sunter automotive cluster on 11 June 1971 under PT Federal Motor. Then in October 2000, the company changed its name to PT Astra Honda Motor (AHM). Honda has been serving Indonesia for more than 38 years and has gained the trust of customers nationwide. It has a production capacity of up to 3,000,000 units per year and employed 13,000 people in December 2007. Along with the increasing demand for motorcycles, Honda gradually increased its production capacity by building the second plant in Pegangsaan automotive cluster in 1996 and the third plant in Cikarang automotive cluster in 2005. With the three plants, along with skilful employees and sophisticated production technology, Honda is able to produce 3 million motorcycles every year (Honda 2004, 2007a; MOTIRI 2005).

Honda in Indonesia has three major plants, two training centre facilities for motorcycle product development and parts development, and one dies and mould division centre. Honda has deliberately located its plants in different automotive clusters across the Java region (Honda 2007b).

The first plant was located in the Sunter automotive cluster (Jakarta) alongside the head office. The second plant was situated in Pegangsaan Dua automotive cluster (Jakarta) as part of Honda's secondary expansion in the Jakarta area. The first and the second plants of Honda Indonesia were the effect of the government regulation to pool all the automotive makers into the Sunter automotive cluster in the 1990s (Kuncoro 1996, 2002).

However, for the third plant, Honda expanded outside the Jakarta automotive cluster, due to the overcrowding of other automotive manufacturers in Sunter and Honda's own internal strategy to be close to its suppliers. Honda Indonesia emphasises that keeping down operational costs in a factory will have a significant impact on its integrated production system, NH circle (Honda 2007b).

Transportation costs, for example, have to be considered in broader terms than simply the circulation of materials and goods within the factory. They must include the delivery of parts from suppliers and the movement of the products from the plant to the ports at Tanjung Priok-Jakarta. For knockdown parts, costs would also include transport to factories in other countries, unpacking, assembly into finished goods, delivery to dealers and finally to the customer. Only by knowing about all these transportation costs would one have a proper starting point from which to examine costs:

> Previously, Honda's suppliers, both tier 1 and tier 2, were less efficient to supply Honda's plants in Sunter Plant and Pegangsaan Dua. It was due to different and separated production plants. Now, since the third plant was built, Honda can easily control the supply chain, as this new plant has been able to integrate Honda's production system, NH circle, more efficiently. Honda believes that geographical proximity will be more advantageous for business and knowledge transfer process.
>
> (AHM representative, 27 April 2007)

Therefore, Honda Indonesia has found that specifications for motorcycles were linked to the specific conditions of each export market, and this was relevant for reducing distribution costs. Consequently, the third plant was located in Cikarang automotive cluster (West Java), and it has become the most modern Honda plant in Indonesia.

The Indonesian automotive cluster supply chain

The importance of being located in the automotive cluster

Essential to a thorough investigation of the Japanese (or joint venture Japanese-Indonesian) transplants parts sourcing is to examine where the firm uses a direct supplier to obtain their own components and materials. This is particularly important given the emphasis often placed in studies of Japanese automotive industry on the multi-tiered (up to ten levels) nature of automotive supply linkages in Japan (Honda 1991, 2004, 2007a; Mair 1994), which raises the question whether such hierarchies will be reproduced in the Indonesian automotive cluster.

In fact, it is even more difficult to obtain comprehensive information at the second-tier level than at first-tier level due to the confidentiality of Honda Indonesia's purchasing data. However, based on the interviews both with Honda Indonesia and with its first-tier suppliers, it is known that the relationships of the supplier tiers are complex, as described by the Honda Indonesia representative below:

150   *Motorcycle production in Indonesia: Honda*

>Investigating Honda's suppliers is not an easy task. Partly, it is confidential and also, it depends on the specific product project from Honda Indonesia, as in this case, the first-tier supplier can be the second-tier supplier and vice versa.
>
>(AHM representative, 29 May 2007)

Subsequently, to complete the analysis of Honda Indonesia's transplants parts sourcing, some criticism of Honda's existence in the Indonesian automotive cluster (Okada 1983; Tarmidi 2004; Rasiah 2005) and the argument are discussed as follows:

1   **'Screwdriver' plants with local content:** this was correct during the early stage of Honda establishment in Indonesia, but it has not been true since the mid-1990s. The charge of low local content, with import of most parts from Japan, was accurate during the early of years of Honda Indonesia's transplant production (Honda 2007a). Nevertheless, since the mid-1990s there has been a rapid build-up of local sourcing. Therefore, the certainty with which critics argued that low content would be maintained indefinitely was misleading (Honda 2007a):

>The construction of Honda Indonesia's expansion in the second and the third plant in Java was a crucial in this respect, as Honda needs to tap into the local knowledge with the local suppliers.
>
>(AHM representative, 27 April 2007)

2   **Only unsophisticated parts are sourced in Indonesia:** this was true during the early 1980s, but since 1990, it has become increasingly inaccurate.

>When production started, local sourcing of parts did consist largely of unsophisticated parts-material and generic components. The pattern has been gradually changed.
>
>(AHM representative, 29 May 2007)

According to Honda Indonesia, analysis of local content levels at supplier transplants reveals that firms using unsophisticated materials as inputs tend to have higher local content than those using more sophisticated components as inputs (Honda 2007b). However, with increased local content the nature of parts sourced in the automotive cluster in Indonesia has changed.

>Local sourcing has progressed over time from generics (e.g. tyres and batteries), materials (e.g. sheet steel, glass) to Honda-specific smaller metal stampings (for body parts), interior and exterior trim (e.g. moulded plastic, interior panels, rubber parts), major and minor mechanicals (e.g. engines, shock absorbers, brakes), and electrical (e.g. instrument clusters, heaters, wiring).
>
>(PT Menara Terus Makmur, 20 May 2007)

In fact, recent improvements in Honda Indonesia have been in R&D and engineering functions. Honda Indonesia has been advancing its product design to be more eco-friendly due to government instructions to lower carbon emissions. Therefore, Honda's local parts sourcing in the Indonesian automotive cluster (mainly in the Java region) has progressed significantly, beginning with simple parts and improving over time to more and more sophisticated parts (Honda 2007b).

3 **Linkages with Japanese suppliers were also transplanted:** this is correct.

> At least 18 Japanese supplier firms belonging to Honda Group in Indonesia located in the same automotive cluster and nearby Honda Indonesia's plants, in order to contribute towards increasing Honda Indonesia's local content.
> (AHM representative, 27 April 2007)

These firms often supply the same parts to Honda in Japan. In fact, HMC invited Japanese firms to invest in Indonesia as a partner of Honda Indonesia (MOTIRI 2005; Honda 2007b). They agreed and continued to establish supplier transplants through joint ventures with local firms. Additionally, there is no doubt that Honda's *keiretsu* in Japan has been transferred to Indonesia.

4 **Supplier transplants invade Toyota markets as well:** this is correct.

> In fact, due to the strong linkages of the Toyota Group in Indonesia, Honda Indonesia's supplier sent over their output to other automotive firms in Indonesia, mostly to Toyota.
> (AHM representative, 27 April 2007)

The supplier also confirmed this statement:

> As a supplier, we cannot solely rely on Honda; we must respond to any order from any automakers to make profit. It is a common thing in dealing with the Japanese automakers. Thus, first tier of Honda Indonesia could be a second tier of Toyota and second tier of Toyota could be a first tier of Honda. Depending on the product, design.
> (PT Aisin Takaoka Indonesia, 30 April 2007)

5 **Supplier transplants exploit joint ventures with domestic firms:** this is correct. In fact, the joint ventures involving domestic companies and Japanese firms have become increasingly important in Honda Indonesia transplants (Honda 2007b). Both the suppliers and Honda Indonesia mutually benefit from this partnership, as PT OTICS Indonesia explained:

> The knowledge transfer between OTICS and Honda has been helpful to improve the production system. We have learnt a lot from Honda and we

can share the engineering problem in Honda Training Centre located in this automotive cluster with the other suppliers.

(PT OTICS Indonesia, 10 May 2007)

As a result, during the early production years in Indonesia, Honda parts sourcing did fit some of the patterns identified by critics (Okada 1983; Tarmidi 2004; Rasiah 2005) on low local content and a bias in local sourcing towards low-cost, low-technology, unsophisticated products.

Furthermore, from Honda's strategic perspective, in the first years of transplant assembly, the first local supplier infrastructure created by a Japanese firm in Indonesia had to be built slowly and carefully in an exploratory fashion. This involved both tentative links to domestic firms and the establishment of a small number of transplant suppliers in vital areas.

> Links to domestic firms turned out to be fraught with difficulties related to manufacturing process and product quality. Therefore, establishment of suppliers' transplants in the same cluster with Honda, even though requiring the first global outposts of many medium-sized and small manufacturers with no experience outside Japan, emerged more successfully.
>
> (AHM representative, 27 April 2007)

In fact, a very large percentage of Honda's transplant suppliers are owned by the same firms who supply the company in Japan, and Honda maintains a very close working relationship with these firms; many of them have designed and developed their own products and Honda's willingness to lend personnel and sometimes financial support has helped overcome potential difficulties.

Honda's supply value chain network

Honda Indonesia has already built a network of local parts makers and has begun to manufacture engines. The current list of Honda's 18 main suppliers is provided in Table 8.1.

Specific information on Honda Indonesia with domestic suppliers (first tier and second tier) is not easily obtained. This may reflect a strong *keiretsu* within the Honda network. There is some indirect evidence for such an interpretation in: (1) journalistic accounts of Honda Indonesia's domestic suppliers; (2) the tendency to source particular parts preferably from Honda Group as a priority and non-Honda group as an alternative; and (3) the multiplicity of Honda's relations with Japanese supplier transplants which reveals what parts are not supplied by domestic firms

It is nevertheless clear that following its early purchase from domestic suppliers, Honda Indonesia has steadily increased its relationships with domestic firms (Honda 2007b). Indeed, Honda Indonesia maintains that its first priority in

Table 8.1 Honda's supply value chain network, Indonesia

No.	Supplier to Honda
1	PT Aichikiki Autoparts Indonesia
2	PT Asama Indonesia Mfg
4	PT Atsumitec Indonesia
5	PT Best Logistic Service Indonesia
6	PT Honda Lock Indonesia
7	PT Honda Power Products Indonesia
8	PT Honda Precision Parts Manufacturing
9	PT Honda Prospect Motor
10	PT Honda Trading Indonesia
11	PT Indonesia NS
12	PT Indonesia Stanley Electric
13	PT Kaneta Indonesia
14	PT Mitsuba Indonesia
15	PT Molten Aluminium Producer Indonesia
16	PT Toyo Denso Indonesia
17	PT Tuzuki and Asama Mfg
18	PT Yutaka Manufacturing Indonesia

Source: author's fieldwork (2007).

sourcing for Indonesian production has always been to find domestic suppliers, followed by, in order:

1 Joint ventures or technical licensing arrangements between domestic firms and established (i.e. Japanese) Honda suppliers
2 Established suppliers that construct transplants or in-house production
3 Importing parts from Japan or other countries

Nevertheless, some parts produced by domestic firms are not available to Honda Indonesia, since domestic suppliers of many components have come under the control of the Toyota group, which dominates higher levels of vertical integration in the supply chain than is usual in Japan (**Veloso 200?**; Sato 2001; Andersen and Christensen 2005).

Moreover, Honda Indonesia argues that in the initial years of motorcycle assembly when output volumes were relatively low, many independent domestic firms did not believe it worthwhile (and by implication, lost their chance) to meet its product specification (Honda 2007b). Therefore, while many domestic firms have since submitted to the 12–18 month process of seeking qualification as a Honda supplier, less than 5 per cent of applicants have been accepted (and acceptance does not necessarily imply a contract).

As a result, it can be argued that domestic firms are often unable to meet Honda's demands for product quality; hence, domestic firms selected to supply Honda have invariably been required to restructure their entire process of delivery and even their manufacturing processes. Initiative in product design is expected as well. PT ASAMA Indonesia, one of Honda's first-tier suppliers, explains:

> Being a supplier of Honda, specifically as a first-tier supplier, the supplier must pass the rigid process in Honda's standard. Maintaining the quality of production system during this verification process is taking up to six months, while having very intense product design control and engineering visits.
>
> (PT ASAMA Indonesia, 22 May 2007)

Consequently, to improve the performance of shortlisted suppliers, Honda Indonesia encourages those suppliers to share knowledge in its part centres located in the automotive cluster, where Honda's engineers can likewise share their expertise to upgrade supplier performance. Therefore, it is not impossible for suppliers to bring their own knowledge and engage in dialogue with Honda in order to get approval.

This aspect of the Honda learning centre and the transfer of knowledge will be elaborated in the next section.

Knowledge transfer within the Indonesian automotive cluster

Knowledge transfer and Honda's local suppliers

In the mid-1990s, suppliers for Honda Indonesia had solidified their technical capabilities, despite economic turmoil, to furnish a wide variety of components including instrument clusters, wiring harnesses, hose tubes, and interior and exterior trims. Of prime importance among the components more recently produced in the automotive cluster in the Java region are electrical parts and mechanical parts (MOTIRI 2005; Honda 2007b). Suppliers are also involved in a whole range of products including the most complex high-value parts.

From Honda's strategic perspective, in the first years of assembly transplants, the first local supplier infrastructure created by a Japanese firm in Indonesia had to be carefully explored and slowly built (Honda 2007b). This involved both tentative links to domestic firms and the establishment of a small number of transplant suppliers in vital areas. Links to domestic firms turned out to be fraught with difficulties related to the manufacturing process and product quality. The establishment of transplanted suppliers was tough, requiring many small- and medium-sized manufacturers with no experience outside Japan to establish their first global outposts, but these emerged more successfully.

In Honda part centres where Honda's engineers can share their expertise to upgrade the supplier's performance, the suppliers also bring their own knowledge and have discussions with Honda. Ongoing engineering visits like these have been supported by the proximity between Honda Indonesia and its suppliers.

Based on the interviews with selected industrial developers where Honda's plants are located, there is a tendency for the Honda supplier network to be located either in the same area or nearby:

> As the industrial estate developer, we are pleased to see the tendency in Honda case. Before Honda actually moved into this cluster, its suppliers had already set up an agreement to expand their factories in the same location as Honda located. Therefore, when physically Honda built its third plant, its suppliers followed what Honda did. However, for those suppliers who are not able to make such expansion, they are hoping to follow their counterparts soon.
>
> (MM2100 Industrial Estate Developer, 10 May 2007)

In general, Honda, unlike Toyota, has never established a formal association to group its suppliers (Honda 2004). In fact, a very large percentage of Honda's transplant suppliers are owned by the same firms that supply Honda in Japan (Honda 2004, 2007a). At Honda, opportunities for its workers and suppliers are developed in a variety of ways, notably through NH circle activities mostly based in the Honda Indonesia training centre located in the Sunter automotive cluster in Jakarta (Honda 2007b).

> If the training is focusing on the development of parts, then it will be held in Honda Indonesia parts centre in Cakung-Jakarta. Representative from suppliers (both tier 1 and tier 2), dealers, and customer are always attending this training. But, for particular queries for dies manufacturing, the suppliers and Honda Indonesia usually discuss in dies manufacturing division located in the first plant.
>
> (AHM representative, 27 April 2007)

The NH circle activities used in Honda's educational aspect is the Honda counterpart of the quality control (QC) circles of many Japanese automotive companies (Honda 2004, 2007a). The NH circle system now operates in the entire Honda group as well as in the companies and workshops associated with each local factory. In the mid-1980s, NH circles were set up in Honda companies overseas where they have produced remarkable results (Mair 1994; Honda 2004).

Honda created the term 'NH circle' – N stands for now, new or next, and H for Honda – with a view to taking a broader approach.

> Honda rejected the QC circle because it dealt only with cost reduction and quality control. NH circle allows members to bring up any problem and provide a venue for discussing a wide range of subjects. This system allows workers to improve their interpersonal relationships and promotes closer teamwork, and it encourages them to raise questions of their own and to work together to find the answers.
>
> (AHM representative, 27 April 2007)

Subsequently, NH circle activities are voluntary and work-initiated; they help workers (as well as suppliers) to improve their work procedures and to generate real interest in what they do. However, an NH circle is not just about the activities aimed at promoting bottom-up movement in the corporate organisation, but also about improved top-down communication and understanding.

> Honda's NH has more freedom than QC groups in other Japanese automotive companies, and includes the element of play in their activities. Honda is known for its 'Idea Contest', which is open to all its employees.
>
> (AHM representative, 29 May 2007)

In addition, the NH circle and the Idea Contest along with the expert system, on-the job training (OJT) and other training programmes, the overseas trainee system and the round table meetings for top management and other executives, are all ways of ensuring lively two-way communication at Honda (Honda 2007a). Accordingly, there is a 'virtuous cycle' in which information from the bottom flows to the top and then passes back down in the form of proposals, recommendations, and policies.

The implementation of NH circles in Honda Indonesia was extended to small-group activities of the workers themselves (Honda 2007b). These activities and the concept of total quality control are in tune with basic human impulses and patterns of organisation.

> NH circle activities in Honda Indonesia focus on the following four points. First, incentive or motivation is vital. Second, rewards should be clearly specified: who achieved what and how the achievement was rewarded. Third, people delight in festivals, and NH circle activities should be a release from routine. Fourth, the pursuit of efficiency through NH circle activities should not lead too easily to the transfer or dismissal of workers.
>
> (AHM representative, 29 May 2007

Nevertheless, the NH circle is not a fail-safe remedy for management. Its effectiveness depends on whether or not the corporation has a management philosophy of its own and well thought out strategies for market competition. Also, it depends on whether or not the company has developed a distinctive style and tradition of management.

Accordingly, Honda does have distinctive systems of management and production (as does Toyota with its TPS). Only when the company has its own philosophy and the ability to create its own management and production systems can total quality control activities be used to strengthen those systems.

Honda management is not precisely the same in every country; instead, it is tailored to meet specific needs and conditions (Honda 2007a). There is nothing surprising about this, as Honda wants to pay attention primarily to individual differences. Thus, Honda management is a mix of many elements drawn not only from Japanese experience but also from overseas experience, including the response of users in the host countries and local government expertise. Therefore, Honda's production and management system is a product of its global experience, including the valuable lessons learned in Honda Indonesia.

The trust-based relationships Honda has established with thousands of suppliers around the world have become crucial to maintaining stable production and fulfilling the commitment to the continuing enhancement of quality and advanced product functionality. Recognising the importance of the relationship with its suppliers, Honda is building long-term relationships.

Accordingly, in striving for growth through long-term relationships, Honda's purchasing division takes care to provide equal opportunity to any supplier who seeks to do business with them. Honda Indonesia chooses suppliers via fair processes while respecting their independence and treating them as equals (Honda 2007a, 2007b).

> When purchasing parts and materials, we select a business partner by impartially comparing and evaluating various candidates based on technological strength, product quality, timeliness of delivery, cost, financial state, regulatory compliance, environmental record, handling of confidential information, and other factors. Contracts with suppliers are based on requirements of compliance with prevailing laws and regulations.
>
> (AHM representative, 27 April 2007)

Faced with increasingly diverse customer needs and rising expectations of product quality, Honda Indonesia depends on its strong partnerships with suppliers to deliver products with superior QCD (quality, cost, delivery).

By being located in the cluster, Honda Indonesia is able to disseminate the production knowledge and to work closely with the suppliers through NH circles, from the initial stages of product development, exchanging opinions and information to enhance safety, functionality, environmental performance and other factors. For example, representatives from the purchasing department of Honda Indonesia will visit suppliers' factories and inspect production processes to ensure that the need for a stable supply of high-quality products at a reasonable cost is fully satisfied.

Revo Project: Honda Indonesia's localisation project

Honda Indonesia launched New Revo, which has a more stylish and sporty appearance and design on Friday 29 February 2008 as a localisation project from Indonesian engineers. According to Honda's global-local business strategy, the need to produce local motorcycle designs is necessary for penetrating local markets.

The Indonesian engineers in Honda Indonesia initiated this project and approached the parent company in Japan, which demonstrates that the process of knowledge transfer in Honda Indonesia has indeed been gradually progressing. In fact, this project is located in the third plant in the Cikarang cluster where Honda is closer to its suppliers.

New Revo 2008 is the pride of Indonesian engineers. It appears in more elegance, luxurious, futuristic, dynamic and trendy performance. The bodyline gives sharp shades, modern and sporty. The synergic of slender and sharp body creates aerodynamic shapes. This localisation project is a pride of the Indonesian engineers.

(AHM representative, 29 May 2007)

Despite having an overtly stylish and sporty design, New Revo 2008 is also practical and responsive, with manoeuvrability as a key feature. The position of an ergonomic steering handle makes Revo more comfortable, nimble and stable to ride.

Since it was launched in April 2007, Honda Revo has gained a reputation as the 'Best Seller Motorcycle 2007', and sold 56,275 units in just nine months. This prestigious achievement demonstrates Honda Revo was well received by the local market.

Links between Honda Indonesia and the university

In terms of knowledge transfer in the wider community, with a particular focus on automotive engineering, Honda Indonesia, with Astra Indonesia as a business partner, has been active in several educational projects. This is a usual channel for an FDI company in Indonesia to contribute benefits to the host country, in response to both government regulation and business contracts, and the company's own business philosophy.

Subsequently, Honda Indonesia is developing an automotive engineering programme in university, mostly in the Java region. Under the Astra Group, the Indonesian automotive conglomerate, Honda has been extending its knowledge to the university, by getting involved in providing engineering visits to Polman-Politeknik Manufaktur Astra (Astra Manufacturing Polytechnic) and engaging in collaborations with the university across the Java region as part of their industrial and engineering laboratory tests, based on product project development. Additionally, Honda Indonesia also opens its transplants for factory visits, to serve as a teaching factory for the engineering faculty of the university across Indonesia.

Nevertheless, gaining detailed information about how strong Honda Indonesia's links are with the university is not an easy task as Honda Indonesia also has a car division. The Honda car division has stronger links to the university and contributes to the development of the automotive dissemination agenda, for instance guest lecture presentations, workshops, and OJT programmes in certain engineering faculties across the university in the Java region.

Conclusion

Honda plants are located in three different clusters in the Java region along with three different support centres. The decision to be located in three automotive clusters in Java has made Honda Indonesia an example of how a motorcycle manufacturer can transfer knowledge and technology to the local economy, particularly through supplier networks.

The building of motorcycle plants and support centres by Honda Indonesia in the automotive cluster has accelerated the growth of the cluster. In fact, Honda Indonesia's external linkages (international strategic alliances and the development of new supply chain structures) and internal linkages (team working, network management structures, flexibility, etc.) have been expanding simultaneously with its leading position in the Indonesian motorcycle market.

In relation to the transferability of Japanese management techniques, Honda in Japan has transformed Honda Indonesia away from 'screwdriver' actions, by implementing NH circles in Honda Indonesia. This learning dimension is supported by Honda's geographical proximity to its suppliers in the cluster. Therefore, being located in the automotive cluster is important for Honda Indonesia to expand its knowledge and technology within the company itself and with suppliers. Honda Indonesia has its own technical capacity building by having knowledge transfer within the Honda group and among the suppliers. Additionally, in developing indigenous initiatives through NH circles, Honda Indonesia launched the New Revo Project as a localisation project. This project has been launched successfully and it has fulfilled customer needs and demands.

Additionally, Honda Indonesia is also concerned with the development of an automotive engineering programme in the Indonesian University, mostly in the Java region. Therefore, under the Indonesian automotive conglomerate the Astra Group, Honda is involved in providing engineering visits to Polman-Politeknik Manufaktur Astra, although this programme is more linked to Honda's car division than its motorcycle division.

9 The Indonesian automotive cluster in Toyota and Honda's global production networks

In Chapters 7 and 8, I sought to answer the two research questions concerning outcomes and process (see Chapter 1).

Following on from these, in Chapter 9 I provide an analysis of the Indonesian automotive cluster and the global linkages by using production and supplier networks from Toyota and Honda. Therefore, in the first half of this chapter, I purposely explore the importance of Toyota and Honda in the Indonesian automotive cluster as prime movers for knowledge transfer, as a complementary analysis to the case studies in Chapters 7 and 8. In conjunction with that, in the second half of chapter I intend to analyse the research question relating to content: how has the global-local network in the automotive industry impacted on the automotive cluster in Java? I do this by looking at the importance of fostering global linkages for the Indonesian automotive cluster, mainly with the ASEAN-4 (Indonesia, Thailand, Malaysia and the Philippines), where global automotive networks have underpinned an important development strategy.

Finally, I will respond to the research question concerning recommendations: how can the cluster policy be improved? And what are the lessons and recommendations for the Indonesian automotive cluster with regard to sustained development? This issue is mainly dependent on the big players like Toyota and Honda for the sustainable development of the cluster.

Toyota and Honda's significant presence in the development of the Indonesian automotive cluster

Toyota and Honda serve as a global-local transmission vehicle for the best practices of investment in the automotive cluster in the Java region. In fact, Toyota and Honda in Indonesia have been using their Asian production alliances partly as a platform from which to continue supplying high-technology products to Western markets (TMI 2007, Honda 2007a).

The presence of Honda and Toyota is significant as technological knowledge in the automotive industry has a tacit nature and cannot be codified through plans and instructions (Chen 1996; Larsson 1999; Dyer and Nobeoka 2000; West 2000; Fujimoto 2007). This type of knowledge can only be learned through everyday practice on the shop floor and the use of technology. This is a highly relevant issue, particularly in the early stages of the technological life cycle. In these conditions, the use and transfer of new and non-codified knowledge is the key to successful development, particularly with regard to new product design (Fujimoto 2007). Therefore, the presence of Toyota and Honda in the Indonesian automotive cluster is important as technological development is ongoing. The importance of knowledge spillovers and information sharing in production activities leads to localised networking of the automotive firms engaged in related product development activity and to proximity to the sources of the prime movers (i.e. Toyota and Honda).

Additionally, it must be noted that the supplier networks (i.e. vertical linkages) maintained by Toyota and Honda in the Indonesian automotive cluster cannot help but be exclusionary. Interestingly, US and European multinational enterprises (MNEs) in the automotive industry will face challenges entering the Japanese market and production base in Asia (including Indonesia), as it is dominated by *keiretsu* relationships in the Asian automotive clusters (including Indonesia). Breaking into the Asian market, therefore, will become more difficult as the market share and power of such networks increase (Poon *et al.* 2006).

Western automotive manufacturers will experience hard times in this industry as Japanese MNEs are fully supported by the Japanese government (although in the early days of overseas operations, Honda was the only self-reliant company). They have established the most extensive vertical *keiretsu* networks in the Asian automotive clusters, including the Indonesian automotive cluster, and have used these networks to exploit their technological advantages and expand their already considerable claim on world markets. Western high-tech firms, on the other hand, generally do not use long-term contracts with suppliers and do not recognise their operations under the principle of vertical quasi-integration (West 2000; Ozawa 2005). This puts them at a disadvantage.

The existence and importance of Toyota and Honda in the automotive cluster in the Java region has generated much interest. The continuing interest of the automotive MNEs in the Indonesian automotive sector is perhaps due to the way in which it is seen as a paradigm of wider social, geographical, political, and economic phenomena. In Indonesia, a combination of targeted industrial policy (i.e. Indonesia 2025 vision) and particular national conditions (i.e. manufacturing conditions in the automotive industry) has attracted MNEs to this sector. In an attempt to build and bring in the benefits from 'a knowledge-based economy', the development of a high-technology sector such as the automotive industry has been the predominant intention of policymakers in a developing country such as Indonesia. The transfer of knowledge which can come from Toyota and Honda has the potential to result in the creation of a clustering process in the automotive industry in the Java region, which goes beyond that of a global value chain system.

Additionally, given that Japanese automotive MNEs in the automotive industry were initially attracted to Indonesia due to industrial policy as well as the intent of tapping into local capabilities and a local market, it may be expected that such investment organisations will later on engage in global linkages.

162 *The Indonesian cluster in global networks*

Accordingly, Figure 9.1 indicates the link between Toyota and Honda in the Indonesian automotive cluster and the global environment in terms of networks with suppliers, customers, and competitors. Furthermore, the importance of the Indonesian subsidiaries to the parent company (i.e. Toyota and Honda) is applied to an indigenous activity known as a localisation project, which is counted as a significant contribution to the host country. Figure 9.1 is a framework describing the research analysis before fieldwork.

Subsequently, based on the empirical case studies of Toyota in Chapter 7 and Honda in Chapter 8, it has been shown that industrial activity in the automotive sector has been progressing gradually, as the reciprocal benefits from MNEs to local economies and vice versa develop, such as access to resources, technological capabilities and knowledge spillovers. Figure 9.2 shows the revised research framework following this fieldwork. Empirically, Toyota and Honda are more likely to engage in formal technological cooperation with firms in the Indonesian automotive cluster (particularly with the suppliers) than those

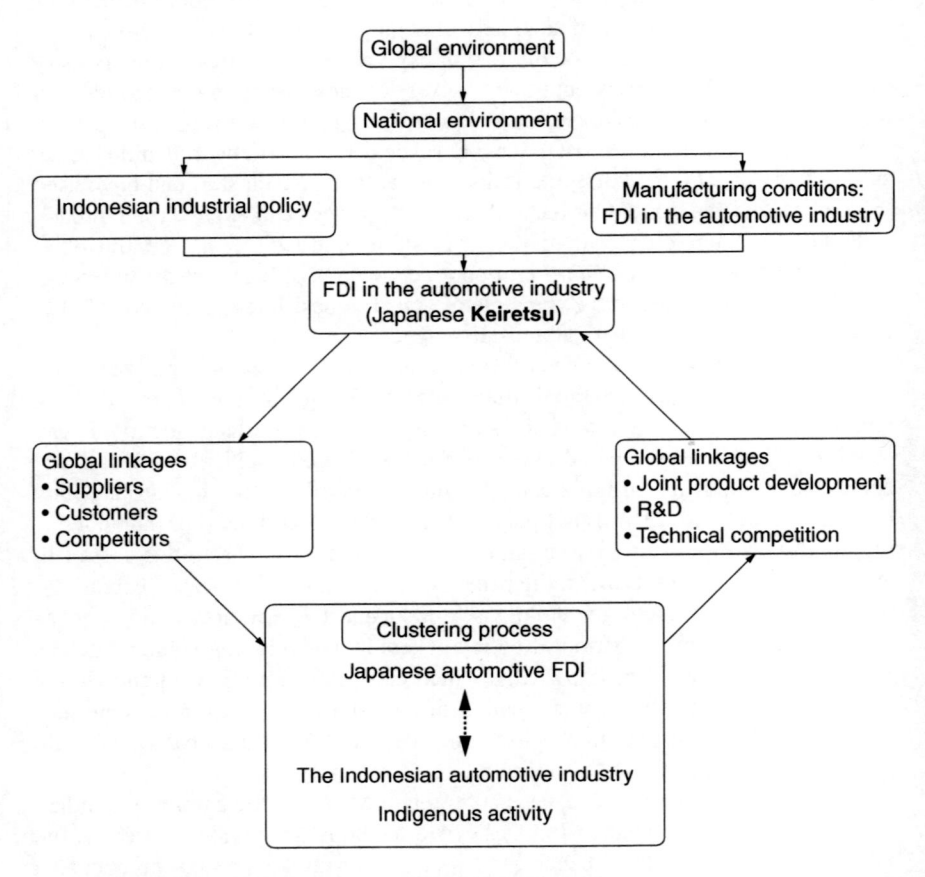

Figure 9.1 Research framework analysis before fieldwork.

The Indonesian cluster in global networks 163

affiliates not located in the cluster. These findings fill a significant gap in the literature as mentioned in Chapter 2 in studying clusters from the perspective of developing countries, with specific attention to the automotive sector.

The case studies of Toyota and Honda found that knowledge spillovers through knowledge transfer tend to be geographically constrained and that MNEs are responsive to agglomeration potential in the industrial cluster in the Java region. The presence of both intra- and inter-firm spillovers was found to be important for attracting the technological activities of MNEs. The case study

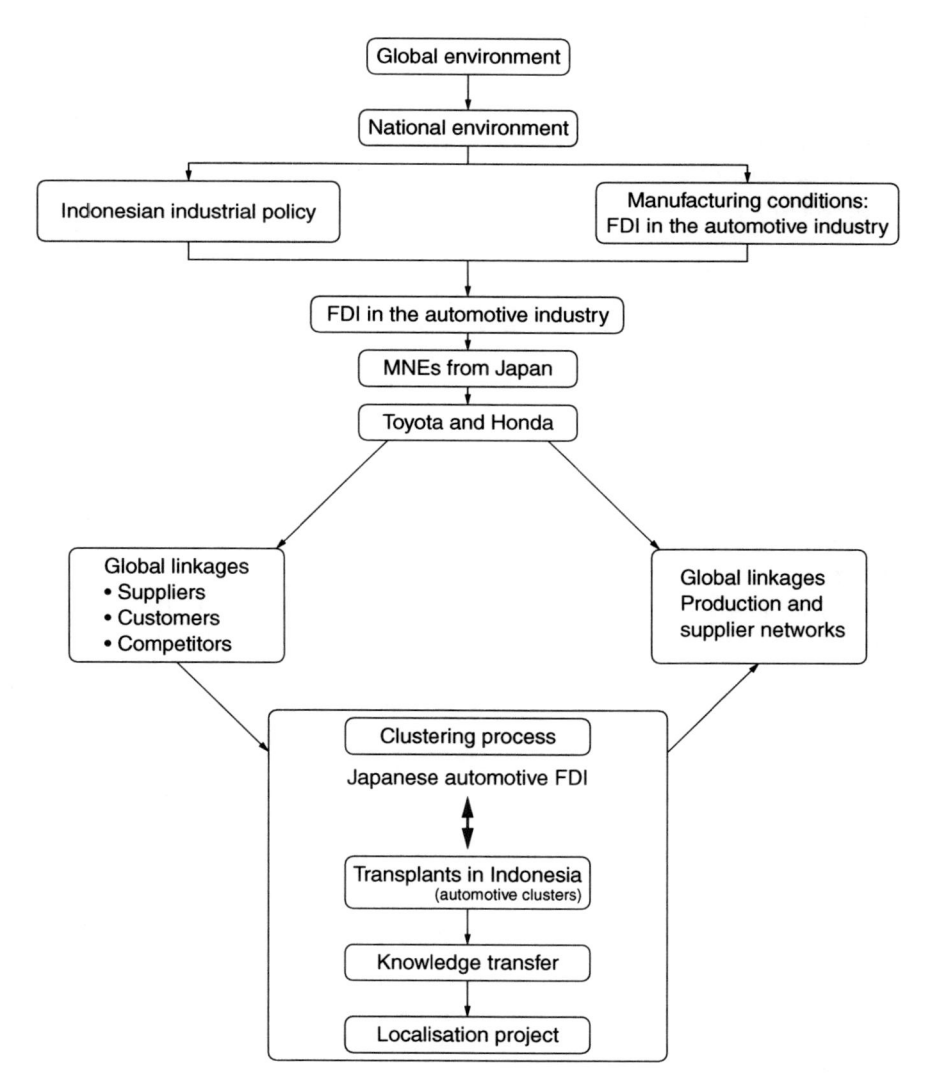

Figure 9.2 Research framework analysis after fieldwork.

findings support the literature reviewed in Chapter 3 which suggests that the Japanese automotive MNEs have heavily relied on their *keiretsu* in terms of transferring knowledge both to their subsidiaries in the host country and to suppliers. Additionally, the case studies in Chapters 7 and 8 complement the literature in Chapter 3 on Japanese automotive foreign direct investment (FDI) and its supply chains in the ASEAN-4 automotive clusters.

Moreover, in terms of the contribution of the Japanese automotive MNE to the cluster, the case studies have shown that the presence of Toyota and Honda has been driving the catching-up process of the technology in the automotive industry, as well as enhancing the degree of internationalisation of local firms to expand the network of the automotive cluster in different locations, both in the local and globally. In addition, the empirical work has shown how Toyota and Honda can generate cluster advancement as well as cluster dynamics, specifically through the transfer of technology and by enhancing the reputation of the Indonesian cluster as regards car and motorcycle production in the global automotive network.

Nevertheless, the degree of contribution often has conditions attached, which relate to the characteristics of the Indonesian automotive cluster. Gains for domestic firms are found to occur when a cluster already exists. Similarly, technology transfer from Toyota and Honda can be a good mechanism to advance the Java region, however, certain regional characteristics need to be present and supportive, such as high local capabilities, universities engaging in industry-related research, and specialised workers.

Furthermore, the need for existing indigenous localisation projects to contribute to the Indonesian automotive cluster is regarded as an important factor. In reality Japanese automakers have been making progress towards 'localisation' by purchasing more of their parts from the local suppliers in the automotive cluster in Indonesia since the 1980s (MOTIRI 2005).

This pattern has changed in order to accelerate knowledge transfer processes for the automotive industry in the Indonesian automotive cluster. They are no longer continuing to import the most sophisticated auto parts and electronic components from Japan. In fact, they have been transforming local suppliers and have established joint ventures between Japanese affiliated part manufactures with local firms in the Indonesian automotive cluster. Although in the beginning, Japanese affiliates still employed a large number of expatriates in management positions, this has been reduced gradually with the hiring of local managers. Clearly, it is not an easy task for them.

Clusters in the context of the Indonesian automotive industry

The automotive cluster in Indonesia is part of the global production network of car and motorcycle producers. The companies have developed a capacity for innovation in their car and motorcycle plants as part of their global corporate R&D and production systems, although the levels of innovation are not advanced when compared to their parent companies. As a result, over the past two decades, the automotive sector in Indonesia has striven to become more developed than before and has nurtured appealing spillovers within and outside the automotive cluster in the Java region.

Given that the process of automotive clustering in Indonesia involves the interrelatedness of networks among the manufacturers and suppliers, this research also examines the importance of the Japanese *keiretsu* in developing the Indonesian automotive cluster as part of global-local networks in the integrated production system of the automotive industry. This research analyses the importance of MNEs for enhancing local technological capabilities, specifically for a region that has attracted foreign investors to enhance its future wealth-creating capabilities through cost-effective learning. Therefore, the presence of Toyota and Honda has led to empirical work on the effect of automotive investment on the development of cluster in the Java region.

Nevertheless, the dominance of Toyota and Honda can also retard cluster participation in the Indonesian automotive cluster. In these circumstances, automotive firms which are not part of a network with the prime movers will be isolated and may suffer from greater barriers to participation, since knowledge transfer in Toyota and Honda has become the source of technological spillovers. Therefore, the importance of knowledge spillovers do not proceed in isolation; if geographical proximity to these sources is important, then the firms located in strong network areas should benefit from self-reinforcing advantages and will tend to enjoy permanent advantages in profit margins over firms located in weak networks. Consequently, the automotive firms who have strong networks and are located close to prime movers are becoming more dependent upon complementary knowledge from firms other than their own. Thus, synergy stems from the presence of Toyota and Honda networks in the Indonesian automotive cluster. It is a combination of complementarity and the necessity of coping with dependency upon the business environment that are the driving forces behind the development of such cooperative networks.

In the context of the Indonesian automotive cluster, there is a clear development indicating the growth of industrial networks. The interaction between the various actors' networks of production, as well as the interdependency of complementary competencies and knowledge, has played an important role in defining the automotive cluster in Indonesia. Thus, this automotive cluster can be characterised as being networks of production of strongly interdependent automotive firms (including the suppliers), knowledge producing agents (universities and engineering companies), and bridging institutions (industrial estate developers, automotive business networks) linked with each other in a value-adding production chain.

As a result, the analysis of the Indonesian automotive cluster can provide an insight into both knowledge exchange and the networking process. In fact, the networks within the Indonesian automotive clusters are not generally based on explicit agreements, but a process of cooperation and collective learning. The networks are not fully supported by the Indonesian government but instead rely on local networks supported by the big automotive players like Toyota and

Honda. However, in circumstances where public policy in Indonesia has attempted to create industry agglomerations through attracting the Japanese automotive players to a region, the contribution of such investment to the development of a clustering process is not trouble-free and requires rigorous underpinning policy actions. Although it might be the expectation of policy-makers that such investment will lead to local economies and a clustering effect, there is limited research to either support or contradict this statement (Malmberg and Sölvell 2002; Phelps 2008; Pelegrín and Bolancé 2008).

The learning aspect of knowledge networks

As discussed in Chapters 7 and 8, Toyota and Honda have intensified the learning aspect in their plants in the Indonesian automotive cluster. However, can the Japanese manufacturing system and its associated management techniques really contribute to the Indonesian automotive cluster? Although this analysis cannot speak for every Japanese automotive company, the case studies of Toyota and Honda show that the process of global-local technology transfer can be done gradually, despite the constraints of Japanese secrecy in their pace of technology transfer particularly from the early 1980s until the late 1990s (Chen 1996; Fujimoto 2007).

In line with Japanese understanding of knowledge as primarily 'tacit', most of the Japanese automakers segment tacit knowledge into two dimensions (Nonaka and Takeuchi 1995; Harada 2003; Fujimoto 2007). The first is the technical dimension, which encompasses informal and craft elements in the form of know-how. The second is the cognitive dimension consisting of mental models, beliefs, 'credo', and perceptions shaped by the surrounding environment. Thus, Toyota and Honda have come to realise that tacit knowledge cannot be easily communicated since knowledge also embraces ideals, values, and emotion, along with images and symbols.

In the Toyota and Honda case studies, the process of knowledge transfer begins with technical capability. This process involves transferring knowledge through people by on the job training (OJT). For this reason, OJT has been considered Japan's 'inner mechanism of technology transfer' (TMI 2007a, 2007b; Honda 2007a). OJT not only provides technical and administrative knowledge to the employees, but also coaches them on how to develop higher motivation and better discipline, so that the process of never-ending quality improvement (i.e. *kaizen*) can be fulfilled. Unlike European and US companies, which utilise written manuals and detailed job descriptions (Dyer and Nobeoka 2000; Harada 2003), Japanese companies support all their production management methods and technical training through OJT.

Since Indonesia is still treated as a developing country for Japanese technology transfer, it is difficult to measure and understand the process. In part, this is because the term itself is vexingly vague and it is a quiet complex and long-term commitment process. Consequently, this research defines technology transfer as the method, knowledge, and skill used to improve and enhance the production and distribution of goods and services in the automotive industry. As a result, it can be embodied in different forms: the machinery used in production or distribution; the manuals detailing business procedures; or the minds of technicians, engineers, and managers who design and execute those procedures. For this reason, technology transfer might be viewed as the movement of such a method, knowledge, or skill from one country to another.

In view of that, it can be said that for effective technology transfer to occur, local firms as well as the human resources in the host country must be able not only to operate the imported technology, but also to adapt and master it to suit local conditions. Therefore, it needs an understanding of the underlying nature of the imported technology, and thus a mastery of it, for instance, in the case of transferring automotive engineering from Toyota and Honda to subsidiaries in Indonesia. Both Japanese and Indonesian engineers must adapt and incorporate local values with the standardised Toyota Production System (TPS) and the New Honda (NH) circle. This has been done by choosing OJT programmes, an inner mechanism of technology transfer for sharing the knowledge, and mastering the engineering techniques from Japan.

Additionally, the nature of the technology that Japan transfers to advanced industrialised countries is fundamentally different from that of the technology transferred to Indonesia as a developing country. The former largely consists of patented high-level technology, while the latter mainly consists of modernisation experience and skills closely related to standardised production methods. The scope of a typical technology transfer contract of the second type usually covers production, management, and marketing (Sugiyama and Fujimoto 2000; Fujimoto 2007).

In fact, the arrival of the Japanese MNEs in the automotive cluster in Indonesia in general is a rational and entrepreneurial response to changing cost conditions in Japan. Direct investment from Japan has contributed to the economic development of the automotive cluster in Indonesia, not only by promoting capital formation, production and employment, but also upgrading technological capability through technology transfer. It is all part of a chain of unintended benevolence in which Japanese production causes Japanese economic growth, which begets overseas production in the Indonesian cluster (and Asia in general), which triggers technology transfer, which links to local economic growth.

Nevertheless, the subsequent argument for the reluctance of Japanese MNEs in being 'careful' to anticipate the risk of spilling technology has been the case mainly in the electronics industry (Hatch and Yamamura 1996). It is argued that Japanese companies in this industry are slower in the localisation of managerial and technical personnel, slower in promoting them, and slower in training. In fact, they also appear more reluctant to set up design and R&D units in the host countries (Aoki 1987). Even though they constantly transfer old technology to the Asian electronics industry, 'new technology piles up' in Japan. Hence, year after year, the technological gap between Japan and other Asian electronic producers (South Korea, Thailand, Malaysia and Indonesia) widens. This fact is considered to not be of mutual benefit in the electronics industry, because, if this trend continues, the Japanese economy will become even more dominant in this region (Freeman 1987; Samuels 1987).

In view of this, it is assumed that Japanese MNEs in the electronics industry are acting differently compared to those in the automotive industry. In the electronics industry, due to the nature of the industry, Japanese MNEs do not want to spread their technology to firms outside their *keiretsus* (Ernst 1994; see also Hatch and Yamamura 1996). Japan's large, oligarchic firms enjoy what are called relative asymmetries of access to trade and investment opportunities in their own and their partner's countries (Borrus 1992). This asymmetry, created by government policy and business practices, makes Japanese knowledge relatively difficult to appropriate. Ernst (1994) argues that the closed nature of Japanese regional production networks has constrained the opportunities for host country firms to develop their own technological and organisational capabilities, which are necessary for continued upgrading of their production efficiency and product mix.

In contrast, Japanese MNEs in the Indonesian automotive cluster serve as a particularly effective mode of technology transfer (Tarmidi 2004; Fujimoto 2007. Moreover, it is argued that MNEs in this sector are likely to bring about a more effective transfer than other channels since it involves a sustained relationship between the transferers and the transferee (Dicken and Henderson 2003). This assertion is based on the assumption that technology naturally diffuses through the training of local suppliers in the clusters, who may be expected to meet higher standards of quality control, reliability, and speed of delivery and through the training of local managers and technicians, who eventually might move from foreign to local firms, transferring human capital with them. Nevertheless, the arguments about 'the beginning of Japanese FDIs', particularly in developing countries in the ASEAN-4 has changed (ADB 2005, Ozawa 2005). In other words, in investing directly in the ASEAN-4, Japanese automotive MNEs have shown a tendency to place their plant sites in lucrative market countries where their products already dominate, particularly in the case of Indonesia.

In short, this changing pattern of Japanese automotive MNEs encompasses changes in management style, technology level, and organisational structure. However, it seems to be insufficient to explain overall changes in the global strategy of Japanese automotive MNEs in the context of the drastically changing world economy. Therefore, in interpreting recent empirical evidence from the automotive industry, the beginning of Japanese automotive classification must be understood in order to foster a comprehensive understanding of the Indonesian automotive development.

Additionally, the domination by Japanese automotive firms of the Indonesian automotive cluster has triggered a growing number of local suppliers to be used by Japanese auto manufacturers. In this case, it is not possible to uncover this fact by examining the purchasing patterns of individual firms, which leads to annoyance about the rise in the host country's production. In view of this, it is argued that increased local procurement does not necessarily mean increased business opportunities for domestically owned suppliers. This was confirmed in an interview with Indonesian–Japanese auto part makers:

> In the real story, there was an overwhelming tendency from the big name Japanese carmakers in Indonesia to buy parts from Japanese affiliated companies in Indonesia. The move towards procuring parts from the local has progressed only in the form of Japanese parts manufacturers establishing a local production base. We are lucky enough to have strong ties with subcontractor plants and therefore we can be part of their *keiretsu*.
> (PT NTRI representative, 25 May 2007)

Japanese automotive manufacturers simply responded by coaxing long-time suppliers in Japan to follow them into Asia. Therefore, the level of competition between local Japanese alliance firms and purely local firms is tough. The local firms in many cases have been struggling to survive in a tough market to win the supplier contracts from the Japanese car manufactures. Indigenous Indonesian auto part firms have explained that:

> It is like being a stepchild in the Japanese auto family and it is an ongoing concern. Of course, there is fairness in the bidding process of a new car project from Japanese carmakers; however, the preference is always the priority of Indonesian–Japanese firms, not local firms like us. Therefore, we must show our best performance to compete with them; otherwise, we will not survive in this game. Another downside is the ability to enhance technological capability. For them, it is not a difficult case as they belong to Japanese *keiretsu* so that they can have training and upgrading skill as part of a supplier scheme. But for us, we must be updated by ourselves alongside a limited market both locally and internationally.
> (PT Menara Terus Makmur representative, 20 May 2007)

The Japan Overseas Development Agency and Japan Overseas Development Aid has provided training to local managers and technicians for the automotive firms in the Indonesian automotive cluster. They are using 'matchmaking' programmes to promote the transfer of technology and skills. However, such government-supervised matchmaking programmes have failed to bring many Japanese automotive MNEs to the altar. That is because they ignore microeconomic reality as Japanese firms do not meet and immediately get married. They just do not operate that way. They require a long drawn out 'romance' before they enter into any kind of relationship. Honda Indonesia has confirmed this argument as follows:

> Managing the managerial responsibilities to non-Japanese companies can take place only gradually, and the required managerial skills and know-how are built through in-house, on the job training, and work experience.
> (AHM representative, 29 May 2007)

The automakers are trying to build *keiretsu*-like supply networks in the Indonesian cluster and other Asian clusters to promote technical cooperation and improve the quality of locally produced cars.

> To compete against American and European producers, we need to find the way to reduce costs even further by teaming up with other Japanese 'friends'. Along with that, the purpose of producing a joint product is intended to strengthen Japanese market share in Asia.
>
> (TMMIN representative, 8 June 2007)

Therefore, in the quest for efficient supply networks, Japanese automakers in the automotive cluster are teaming up to form what could be considered super *keiretsus*. This cooperation is designed partly to satisfy demands for the ASEAN market and partly to maintain Japanese domination of the local market.

Indonesian government policy and Japan

The Indonesian automotive cluster, along with other ASEAN automotive clusters, played an important role in the industry's early stages which geographically also made Southeast Asia a logical focus for early Japanese automotive expansion. Japan's proximity to, and war-time position as occupier of, Southeast Asia provided contacts and encouraged a view of Southeast Asia as a strategic buffer for Japanese firms. Product compatibility also played a role.

In the Japanese situation, government has penetrated business and business has penetrated government through a process known as 'reciprocal consent' (Perdana and Friawan 2007). In other words, Japan's government–business network is a mutually reinforcing alliance or partnership that is capable of strong, decisive action as long as it keeps to the established, conservative policy line. Furthermore, the Japanese government–business network has followed the line carefully in Asia, particularly in Indonesia and Southeast Asia, which has long been identified as critical to Japan's national security. Consequently, Japan has tried to cultivate close relations with elites in the region, aimed at securing political and social stability, as well as liberal trade and investment policies vital to Japanese capital.

At the beginning, Japan's economic cooperation policy in the Indonesian automotive cluster was based on the need to secure a steady supply of raw materials and low-cost production. However, in the mid-1980s, the foundation underneath that policy shifted when the dramatic appreciation of the yen undermined the international competitiveness of virtually all the manufacturing enterprises that exported from Japan (Ozawa 2005).

As a result, Japanese industry, in particular the automotive exporting industry, began to see the Indonesian automotive cluster as a regional extension of its home base (Terry 2002). Because of that, the government–business network promoted a new vision of Southeast Asia as integral parts of a Greater Japan, hosting critically important links in an expanded Japanese production and exporting alliance (ADB 2005). Furthermore, Japanese government and business alike believe that the globalisation of economic activity made it impossible to push ahead with economic development within the limited framework of a single country defined by strict national boundaries – particularly not in the Asia-Pacific region, which was rapidly developing into 'one large economic zone and centre of the growth' (FAIR 1989; UNCTAD 2007).

Nevertheless, the Japanese government–business network argued against the process by which MNEs establish cheap production facilities in the Indonesian automotive cluster and export manufactured goods back to their home country. In order to avoid this position, they began promoting a new division of labour within the region, based on each nation's technological level, rather than on its resource endowments (ADB 2005). Therefore, by serving as the region's innovation leader, Japan is a powerful R&D machine that will push forward the frontiers of world demand by actively promoting the development of new products and new technologies that differ from existing types of commodities.

Additionally, Japanese automotive MNEs have been helping Indonesia to support industrial development through industrial clusters (mainly for the Japanese automotive makers), ports, airports, roads, rail lines, and the telecommunication system. Therefore, the Japanese and the Indonesian governments have been helping each other to promote foreign investment as well as to arrange joint ventures and technology tie-ups (Doner 1991; MOTIRI 2005; UNCTAD 2007).

In view of this, the strong conviction has taken root that Japanese developmentalism through MNEs in the automotive sector, including Asia (and Indonesia in particular), has generated benefit for this region (Doner 1991; ADB 2005; UNCTAD 2007). Under this 'developmentalism', innovating manufacturers in the automotive industry rapidly increased their productive capacities, turned to exports, and began achieving dynamic technological efficiency (Sugiyama and Fujimoto 2000).

Nevertheless, an additional question concerns the cost of 'Japanese developmentalism'. Cooperation between government and business in both Japan and Indonesia has in recent years turned collusive, even corrosive (MOTIRI 2005; ADB 2005). Ham-fisted bureaucrats, corrupt politicians, and influence-peddling business executives have become commonplace in this business. More importantly, government policies to promote dynamic technological efficiency have led to a dual economic structure, one characterised by technology-intensive, export-oriented firms at the top and a far greater number of labour-intensive, domestic-oriented firms at the bottom. In addition, cooperation between members of Japanese *keiretsus* and Indonesian bureaucrats has proved exclusionary and thus highly controversial (ADB 2005). Despite complaints from sources both inside and outside Japan, 'developmentalism' persists. This is because it is a system of policies and practices now deeply ingrained in the Japanese political economy. It has become a solid structure of incentives that resists change, even though it has outlived its usefulness (Ozawa 2005).

For Indonesian automotive firms, the benefit of developmentalism via quasi-integration is substantial, particularly in the early stage of network formation

(MOTIRI 2005), when these firms receive crucial infusions of capital, technology, and managerial guidance from the Japanese government–business network.

Nevertheless, it is expected that if policy remains unchanged, the Indonesian and Asian automotive firms will become captive companies much like the low-level subcontractors in a vertical *keiretsu* (i.e. supply chains) in Japan (Sugiyama and Fujimoto 2000). In other words, what had once been a beneficial alliance will prove increasingly costly as the automotive sector of Indonesia and Asia find themselves stuck in a subordinate role. Furthermore, since Japanese firms, faced with the prospect of slower growth, have captured the gains of cooperation, they will become even less willing or able to transfer technology (Ozawa 2005).

The Indonesian government has followed the example of Japan in the 1950s and 1960s by reviewing technology agreements before they are signed (Tarmidi 2004; MOTIRI 2005). Although this might discourage some foreign firms from promoting tie-ups in the first place, and thus reduce overall opportunities for technology transfer, Indonesia is convinced that an effective review programme would improve the bargaining position of local firms, allowing them to gain better agreements with fewer restrictive provisions. This initiative has stated that MNEs must transfer their technology and benefit for the local community in comprehensive ways (economy, social-cultural, education, technology). Subsequently, although the Indonesian government has adopted measures designed to promote supporting industries, they have often ended up assisting foreign MNEs to establish domestic facilities rather than domestically owned supply firms (ADB 2005; UNCTAD 2007).

Additionally, there are mechanisms by which Asian (including Indonesian) auto firms may get stuck (Sugiyama and Fujimoto 2000; Perdana and Friawan 2007). This has to do with asset specificity. Since most of the physical and human capital of the subordinate firms is dedicated to maintaining the relationship with dominant parent companies, the subordinate firms are exposed to constant demands regarding price, quality, and time. The parent companies, in other words, are able to squeeze the subordinate firm as it strives to increase its profitability and international competitiveness. Subordinate firms often have little choice but to bow to the pressure if they wish to maintain the value of their assets and continuing benefitting from their ongoing relationship with the dominant partner. This case is indeed reflective of what happened in the case of indigenous auto part companies in the host country who are not part of the Japanese automakers' first or second tiers. The unequal bargaining power among the indigenous auto part firms is a problem of the tight auto industry. If firms do not belong to the Japanese *keiretsu*, it is hard to penetrate the crowded, controlled market.

Moreover, the subordinate firms of the Indonesian automotive companies are reinforced on the government level. Out of necessity the Indonesian government has adopted policies that benefit Japanese capital more than local capital. Indonesia, like other Southeast Asian countries, may be seen as a 'captive economy' in view of the Japanese industrial power.

In this light, there are some concerns to be taken into consideration by the Indonesian government. Indonesia, and other countries in the ASEAN region, must do more to increase their own technological capacities (UNCTAD 2007). This means investing wisely in education, training, and creating stronger links between public research facilities, particularly universities and private industry. One such example is the establishment of the auto-academia in Jakarta, the auto-learning centre for engineers and the auto community.

So far, the efforts of the Indonesian government have been aimed at encouraging the development of companies in the Java automotive cluster as part of the Japanese *keiretsus*. This strategy was largely born out of necessity because, as a developing country, Indonesia did not have the knowledge, skills, and technological capabilities, to upgrade its automotive industry on its own. Even though the catching-up process is not finished, the Indonesian government needs to develop a strategy to advance the cluster beyond the catching-up stage. As argued, catching up can be done largely with the help of the Japanese *keiretsus*.

However, the next step requires more indigenous effort. The Indonesian government, therefore, should take the following steps to facilitate further development of the companies in the Java cluster:

1 Facilitate MNEs in the cluster and its companies; for example, through legislation making it attractive for foreign companies to invest in this cluster. The Indonesian government could also actively seek investors abroad.
2 Encourage innovative collaboration between Indonesian and foreign knowledge centres on the one hand and leading companies from the Java cluster on the other. For example, the Indonesian government could provide venture capital to make it more attractive for companies to engage in potentially risky innovation projects.
3 Set up local Technological Transfer Centres that use government money to help companies upgrade their performance. Such centres could help companies to upgrade their production process and production technologies, but also to help further education levels of the workforce and the managerial and organisational knowledge of the company management.

Conclusion

The automotive cluster in Indonesia represents a global-local network characterised by positive complementarities (i.e. long-term strategic FDI, industrial relations, and intercompany relations). This cluster, which is still developing, has undergone considerable changes since the 1990s due to the globalisation of the world economy and the attraction of FDI.

The exploration of networks in the Indonesian automotive cluster has been focused on the learning dynamic and the contribution of cluster in supporting the catching-up process to enhance technological capability in the automotive industry. In addition, the contribution of Toyota and Honda to the Indonesian automotive cluster has been thought to be mainly in line with that of a host country looking after its own interests with regards to spillovers among the local suppliers, the competitiveness of the local firms, and contributions to the host region and its development.

Based on the case studies of Toyota and Honda, as the main actors in knowledge transfer, the Indonesian automotive cluster appears to have become a good example of the knowledge flow in this country, as well as promoting Indonesian automotive industry competitiveness in the global market.

In view of this, the automotive cluster in Indonesia has provided the following contributions.

First, the location of Toyota and Honda in the Indonesian automotive cluster, which is spreading across the Java region, has provided the automotive cluster with an opportunity for international contact and to catch up with new technology and innovation.

Second, by using the products of indigenous firms, Toyota and Honda enable local firms to expect increased scale of economies which will bring about a further increase in indigenous invention and innovation.

Third, overseas automotive inward investment tends to have multiple effects in the Indonesian automotive cluster through increased inter-industry linkages, either through the attraction of direct investment of component suppliers or through the purchase of labour and intermediate products. In this case, the Toyota and Honda plants have generated substantial effects in the regional economy in Indonesia and surrounding regions (i.e. ASEAN and Asia-Pacific).

In addition, the perspective of forging automotive industrial production networks in the cluster and macro-regional integration is equally valid for the Indonesian automotive industry. This industry, in reality, can function as an engine, transforming a low-wage, labour-intensive, and developing economy into a higher-wage, technology-intensive, and developed economy. On the downside, however, the Indonesian automotive clusters have become perhaps overly dependent on their Japanese patrons.

From the perspective of regional competition, there is a possibility for Toyota and Honda to lose their competitive edge in Asian clusters (i.e. ASEAN-4), as US and European rivals might force Japanese automotive makers to modify or abandon a preferred course of action in the Asian automotive clusters: a sudden, unexpected increase in production costs would cause Japanese automotive MNEs to choose a low-cost but unrelated supplier. These are all possible scenarios in response to forging the automotive industrial production networks in the cluster and macro-regional integration, that is, an attempt from Asian governments to intervene to restrict the behaviour of Japanese MNEs, extricating themselves from Japan's embrace. In fact, to anticipate this, the Japanese government is moving aggressively to secure the nation's innovative capacity in the Asian region. Moreover, Japan is consolidating its system of cooperation that promotes dynamic technological efficiency.

The cluster policy regarding Java's automotive cluster has been successful in forging links between, on the one hand, Japanese automotive makers and local subsidiaries, and on the other hand, between the local subsidiaries and second-tier suppliers located in the cluster.

The networks have been helpful for subsidiary companies to upgrade their performance in several ways: they have received new production technologies from their Japanese patrons, they have improved the skills of their workforce, and they have learned to apply new business concepts, such as quality control and JIT production systems. All of this gives the Indonesian automotive clusters a better chance in the global economy.

What the cluster policy so far has failed to recognise are the noticeable risks associated with the heavy dependence on just two Japanese automotive makers. Therefore, cluster policy needs to start working on a strategy that will link the automotive clusters in Java to other key automotive firms as well. The newly acquired technologies and skills need to be used as an entry ticket to working with new partners.

10 Conclusion

The Indonesian automotive cluster is a conspicuous feature of the global-local production organisation. The cluster policy for the automotive industry in Indonesia can be seen as a good way to boost knowledge and technology in developing countries by focusing on knowledge linkages and interdependencies between actors in networks of production. In response to the growth of the car and motorcycle industry in the ASEAN region and Asian markets, the Indonesian automotive clusters increasingly make efforts to improve their production and management systems, and to contribute to the core of the global industrial system of the automotive, not only in manufacturing and sales but also in technology and management.

Both Toyota and Honda are rooted in their Indonesian subsidiaries: they have developed manuals about production system and production control techniques, and organisational learning has been developed internally by Indonesian staff by referring to Toyota and Honda in Japan, while taking Indonesia's context into account.

This closing chapter comprises my final remarks. I now want to highlight the positive and the negative side of the case studies in correlation with the theoretical perspectives, particularly with regards to the cluster concept and relevant strands in studying clusters.

Additionally, I conclude this chapter with areas for future research, specifically for the Indonesian automotive cluster and clusters in developing countries in general.

The cluster concept

Many definitions of clusters can be found in the literature, representing equally many views on the form, shape, and function of clusters in the development of companies and regions. In this final section, a few words on these issues are appropriate.

In as far as we can agree that clusters are networks of interrelated and interdependent companies, occupying a certain geographical area and supported by an institutional infrastructure (such as government policy), this work has studied a cluster. From a conceptual perspective, this work has used a definition of clusters as an ideal type, that is fully mature clusters with certain characteristics. If clusters do not have all these characteristics, it is because they are not mature clusters.

Put differently, the question as to how to recognise a cluster when we see one must be answered in terms of degrees of conformity to the ideal type. Arguably, few clusters will fully conform to the ideal type as local conditions, industrial differences, social economic history, and culture all affect the form and shape of a cluster. This means that different clusters may mature differently to a certain degree. Although this approach may not be completely satisfactory from a theoretical perspective, it does provide a conceptual framework for empirical research into clusters. From an empirical perspective, this research has demonstrated the value of the cluster concept in understanding the economic development and knowledge transfer of companies and regions. This research has demonstrated that cluster, as defined above, is a valuable concept to:

- understand the dynamics within a network of interrelated and interdependent companies, and to explain how these network dynamics facilitate and constrain the economic development of the companies involved
- understand the role of spatial proximity with regard to the interactions (in particular knowledge transfer) between the companies in the network (cluster)
- understand the role of the institutional context, in particular government policy and the local context dimension with regard to the interactions between companies in the cluster.

More importantly, the cluster concept offers the ability to understand these elements in relation to each other. That is one of the key qualities of the cluster concept: its ambition to understand economic development from multiple perspectives. In doing so, it does justice to the complex reality of everyday life for companies. The conceptual difficulties this implies are outweighed by the in-depth understanding that the cluster concept offers. However, as argued, herein also lies the limitation of the cluster concept: in-depth understanding goes together poorly with the large N-studies necessary to make (statistical) generalisations.

In summary, this research has yielded valuable leads to further the theoretical development of the cluster concept. It has further shown that the cluster concept is a fruitful framework for in-depth empirical research into the kind of geographically concentrated networks of companies we call clusters. Furthermore, cluster concept research offers important lessons for both companies and government policy. The cluster concept, therefore, is an indispensable tool for both researchers and practitioners with regard to the economic development of (networks of) companies and regions.

Theoretical contributions

Several theoretical contributions can be drawn from this research.

1 Clusters in developing countries

With regard to the development of the automotive cluster, the traditional trickle down model of regional development seems to be confirmed. It is a clear case of companies from a higher developed economy (Japan) advancing the development of companies in a developing country (Indonesia) through the knowledge transfer from multinational enterprises (MNEs; Toyota and Honda). Japanese capital and Japanese technological, managerial and organisational knowledge has helped Indonesian subsidiaries and their suppliers to upgrade from simple workbench companies to companies that now export to other countries.

In view of the development of clusters in developing countries, previous research literature (Schmitz 2000; Humphrey and Schmitz 2000) is still relevant, as well as the importance of external sources of knowledge in rejuvenating local knowledge flows that could link local agglomerations such as clusters in developing countries. Nevertheless, it is important for developing clusters to have contingency plans for changes in consumer tastes and demand and for new technologies that might result in market shifts (Pries and Schweer 2004; Garnsey and Longhi 2004). Therefore, it is expected that the clusters in developing countries can be well organised and self aware. Furthermore, it is expected that these clusters can become the mechanism that gathers information, predicts shifts, and finds new opportunities (Breziz and Krugman 1993; Fujita *et al.* 1999; Fujita and Krugman 2004).

Yet, the drawbacks of these traditional development models are equally clear in the case studies of the Indonesian automotive cluster. First, the fruits of the economic development models are enjoyed by a privileged group of companies only, those who belong to Toyota and Honda *keiretsu*. Second, the Indonesian automotive cluster has become a satellite of Toyota and Honda *keiretsu*. While traditional economic development models are capable of encouraging economic development, it is a one-sided and unequal kind of economic growth. Yet, developing countries may have little choice but to follow this trajectory from underpinned to developed economies. The challenge for the Indonesian automotive cluster now lies in finding new economic development models to propel itself on a trajectory that will enable it to take an independent role in the global automotive industry.

2 Clusters and networks in Japanese MNEs in the automotive industry

The Indonesian automotive cluster has developed as an extension of Toyota and Honda *keiretsu*. This means that network relations follow a hierarchical structure, with the Japanese firms at the top as a parent company, followed by the Japanese-owned Indonesian companies, the Indonesian subcontractors within the Indonesia cluster and those outside it.

This structure has enhanced the knowledge transfer process from Toyota and Honda into the Java cluster in several ways. First, the *keiretsu* structure reduced the potential for unintended knowledge spillovers outside the cluster. Since the Indonesian automotive firms are encouraged to do business predominantly within their respective *keiretsu*, the Japanese automotive MNEs as parent companies could be confident that their technological and organisational knowledge would be protected. This encouraged knowledge transfer since the Japanese MNEs would, in the end, enjoy the benefits themselves. Second, the hierarchical structure propagated a shared sense of identity and shared purpose, and a shared set of norms, values, and conventions throughout the *keiretsu*. As argued by, among others, Amin and Cohendet (2004), these soft factors are important for knowledge transfer, in particular with regard to its tacit dimension. An effective transfer of tacit knowledge requires a shared frame of reference, a shared social context to interpret and give meaning to knowledge. The *keiretsu* context has been helpful in this respect.

Nevertheless, it is doubtful whether the dominance of the Japanese MNEs and their *keiretsu* networks will be equally helpful for the Indonesian automotive cluster in the next stage of its development and in the longer term. Accordingly, this would require the Indonesian automotive cluster to start creating knowledge and innovation of its own and consequently to have more engagement with the suppliers outside Toyota's and Honda's *keiretsu*. Knowledge transfer is essential to make use of the experiences and skills that are present in the various companies in the Indonesian automotive cluster and to use them as inputs for the development of new products, services, and processes. Yet, hierarchically organised networks are not a favourable environment for this kind of intensive communication since it requires the leading companies such as Toyota and Honda at the top of the *keiretsu* to relinquish some of their control.

3 Proximity and industrial clusters: the impact on networks and technology

The automotive cluster in Indonesia shows that even in a global economy, spatial proximity is still highly relevant (see Chapter 2 for more on the impact of proximity on networks and technology) (Nonaka and Takeuchi 1995; Nonaka *et al.* 2000; Gertler 2003; Cooke 2006). In the first place, the implementation of the just in time (JIT) production system forges close links between companies in the cluster. From the perspective of logistics, spatial proximity makes the JIT system more effective and much more cost efficient. Also cooperation between the companies involved may benefit from spatial proximity. Although it is possible to transfer tacit knowledge over large distances, spatial proximity enhances the 'social depth' of communication (Morgan 1997a). As argued, knowledge transfer benefits from a shared social context, a shared set of norms, values, and conventions. Spatially concentrated networks of organisations and people can provide such a context (Morgan 1997a). Spatial proximity has helped knowledge transfer within the Indonesian automotive cluster to some extent, such as on the job training (OJT), where engineers from the Japanese companies spend time at the Indonesian companies to transfer their knowledge and vice versa.

Knowledge production through knowledge transfer is no longer perceived as a linear process in this sector, but instead as the result of the complex interaction between numerous actors and institutions (West 2000; Nooteboom 2004; Liu and

Dicken 2006). These actors and institutions and their interconnections constitute a system of highly interdependent agents. This implies that not only actors, but also institutions, play a major role in knowledge transfer.

In summary, this book provides evidence for the assumption that complex activities – be they production-orientated managerial activities such as JIT or the transfer of tacit knowledge – benefit from spatial proximity as it offers more opportunity to practice 'rich' methods of communication, such as face to face conversations.

4 Upgrading of companies in the Indonesian automotive cluster

Being part of Toyota and Honda *keiretsu* has enormously helped the Indonesian automotive companies to upgrade their performance. They developed from work-bench companies doing simple production work to exporting cars and motorcycles globally. This clearly demonstrates the success of the past development trajectory. Yet, as argued, it situates the Indonesian companies in a subordinate position to their Japanese patrons. As long as the Indonesian automotive companies cannot stand on their own feet, this is not a bad thing. However, the next step of development requires the Indonesian companies to take matters in their own hands.

In view of their *keiretsu* philosophy, it is doubtful that either Toyota or Honda will encourage the Indonesian automotive companies in the cluster to further upgrade their competencies to become independent automotive companies. Yet, in the long term, this offers the best opportunity for sustained development for the Indonesian automotive companies. In order to achieve that ambition, the Indonesian automotive companies must learn to work together more and develop innovative capabilities of their own. To be able to develop new products, services, and processes would make the Indonesian automotive companies more interesting partners, not only for their current Japanese partners, but also to other automotive manufacturers. This would necessitate the automotive cluster in Indonesia to assume more of the characteristics of a real cluster, rather than two separate networks belonging to the *keiretsu* of Toyota and Honda. Ideally, the most developed automotive firms in the Indonesian cluster should take the lead in the next step of the upgrading process and initiate innovation projects wherein they combine the knowledge and expertise of several firms in the cluster.

The Toyota and Honda case studies are examples of successful MNEs and knowledge transfer process in the cluster, where companies in a developing country have grown into internationally exporting companies. However, this has resulted in the creation of Japanese dominated hierarchical networks that place the Indonesian companies in a dependent position. If the Indonesian automotive cluster is to develop into a 'real' (or mature) cluster, it has to forge horizontal collaboration between the companies in the cluster aimed at innovation. This process needs to be supported by a new kind of development policy from the Indonesian government.

Areas for future research

Possibilities for future research are in line with the key theoretical contributions of this book, as discussed in the previous section. Obviously, the conclusions give rise to new questions on how clusters mature.

1   To what extent is the Indonesian automotive cluster unique or is it an example of a non-mature/developing cluster in general? And what lessons can be drawn from that for the maturing of the Indonesian automotive cluster?
2   To what extent are new regional development concepts, such as 'sectoral system of innovation' and 'the learning region' helpful in explaining the development of the automotive cluster in Indonesia? The Java region is certainly not yet a learning region in the same sense as the cluster in Silicon Valley. The Java region lacks institutional thickness and the presence of a substantial number of advanced companies. The implication of this is twofold. On the one hand, the 'new' concepts of regional development can be seen as the ideal situation for Java, which means that the question is: what does Java lack to become a learning region and how can it remedy this? On the other hand, it may be that these 'new' models are not well-equipped to study development in developing regions like Java. Therefore, the question may be: what makes these 'new' concepts unsuitable for developing regions and what concepts of regional development are better placed to explain development in these regions?
3   Spatial proximity is argued to be important in the development of the Indonesian automotive cluster. However, more research is needed to explain how and why. Mature companies not only have strong links within the cluster, but also to the outside world. This is necessary to provide the cluster with much needed new knowledge (Morgan and Nauwelaers 2003). Nevertheless, spatial proximity may still be necessary to turn the knowledge acquired outside the cluster into new products and services. Evidence suggests that face-to-face communication and trustful relations still matter in this process of knowledge creation. Spatially proximate relations are argued to be better vehicles for this than geographically dispersed ones. '...To the extent that innovation is based on new ideas and that the creation of new ideas is a social process involving discussion, then geographical proximity is important in innovation' (Best 1990: 235).

   Opposed to this view is the idea that knowledge is exchanged throughout communities of practice (Amin and Cohendet 2004), with a shared frame of reference to interpret and understand knowledge on the one hand, and long-term trustful relations on the other hand, rather than spatial proximity supporting the exchange and creation of knowledge. Therefore, the question is: what mixtures of spatially proximate and spatially dispersed relations best further the development of a cluster?

4   The Indonesian government is certainly not the only one to adopt cluster policies to further (regional) economic development. What lessons can the Indonesian government learn from cluster policies elsewhere? For example, since the 1990s the EU has experience of regional economic development policy where furthering innovation in clusters plays a prominent role. The European experience proves that developing a supporting institutional infrastructure of government agencies, innovation support centres, and knowledge transfer centres is vital for cluster development.

Notes

3 Production and knowledge transfer in Japanese automotive networks

1 *Keiretsu* are a collection of firms producing a wide range of products and centred on a bank and trading company. Each member company owns a small proportion of shares of the other companies. *Keiretsu* has been able to develop auto firms because of the role of the *keiretsu* bank providing the necessary funds and partly due to the advantages of mutual ownership and cooperation in providing assistance with technological change and management systems. Nine Japanese car corporations have emerged as a result of *keiretsu*: Toyota, Nissan, Mitsubishi (24 per cent Chrysler shareholding), Mazda (25 per cent Ford shareholding), Honda, Isuzu (34 per cent GM shareholding), Suzuki (5 per cent GM shareholding), Daihatsu (important Toyota shareholding) and Subaru (important Nissan shareholding).

4 Global and national environment: the macroeconomic context in Indonesia

1 International organisations such as the Asia Foundation and the German Technical Cooperation (GTZ), have also been working with local jurisdiction to improve the performance of OSSs.
2 The DSB panel concluded that the local content requirements of the 1993 and of the February 1996 car programmes were linked to: (1) sales tax benefits on finished motor vehicles incorporating a certain percentage value of domestic products or on national cars; and (2) customs duty benefits for imported parts and components used in finished motor vehicles incorporating a certain percentage value of domestic products.
3 Additionally, the second IMF reform package was formulated in the form of the Letter of Intent signed on 15 January 1998 by former President Soeharto. The reform programme consisted of a 50-point Memorandum of Economic and Financial Policies attached to the Letter of Intent. The impact on the automotive industry was that the government had to discontinue special tax, customs, and credit privileges to the National Car Project.

6 The importance of the Java region for the Indonesian automotive cluster

1 Jabotabek is the metropolitan area consisting of Jakarta, Bogor, Tangerang and Bekasi.
2 Formally modelling the net cost of dealing with the bureaucracy is challenging, particularly as such costs may be endogenous to a firm within a given location.
3 Following Marshall (1890), Arrow (1962) and Romer (1986).

184 Notes

4 The numbers applied to carmakers in Table 6.1 relate to the locations shown in Map 6.1.
5 The numbers applied to motorcycle makers in Table 6.2 relate to the locations shown in Map 6.1.

7 Case study of car production in Indonesia: the Toyota complex

1 *Jishuken* is a voluntary learning group within Toyota and its suppliers. The idea to share knowledge among the engineers is endorsed in this learning group.
2 Toyota's founder Taiichi Ohno said in his 1988 book *Toyota Production System: Beyond Large-Scale Production*, 'We don't just build cars, we build people'.
3 Polman (2008) *Polman Astra*. Available at: www.polman.ac.id.

Bibliography

ADB (2005) *The Key Indicators of Developing Asian and Pacific Countries*. Jakarta: Asian Development Bank.

ADB (2006) *Indonesia: Country and Strategy Programme*. Manila: Asian Development Bank.

Ahuja, G. (2000) Collaboration networks, structural holes and innovation: a longitudinal study, *Administrative Science Quarterly*, 45, pp. 425–455.

AISI (2008) *AISI: Indonesian Association Motorcycle Industry* Available at: www.aisi.or.id (Accessed: 12 November 2008).

Altenburg, T. and Meyer-Stamer, J. (1999) How to promote clusters: policy experiences from Latin America, *Word Development*, 27.

Alvstam, C.G. and Schamp, E.W. (2005) *Linking Industries across the World*. Hants: Ashgate.

Amin, A. (2002) Spatialities of globalization, *Environment and Planning*, 34, pp. 385–399.

Amin, A. and Cohendet, P. (2004) *Architecture of Knowledge. Firms, Capabilities and Communities*. Oxford: Oxford University Press.

Amin, A. and Thrift, N. (1995) Globalisation, Institutional Thickness and The Local Economy, in Healey, P., Cameron, S., Davoudi, S., Graham, S. and Madani-pour, A. (eds) *Managing Cities: The New Urban Context*. London: John Wiley and Sons, pp. 91–108.

Andersen, P.H. and Christensen, P.R. (2005) Bridges over troubled waters: suppliers as connective nodes in global supply networks, *Journal of Business Research*, 58, pp. 1261–1273.

Andersson, U., Forsgen, M. and Holm, U. (2001) *Doing Critical Management Research*. London: Sage.

Aoki, M. (1987) The Japanese Firm in Transition, in Yamamura, K. (ed.) *Domestic Transformation*. Stanford: Stanford University Press.

Aoki, M. (1988) *Information, Incentives, and Bargaining in the Japanese Economy*. Cambridge: Cambridge University Press.

Ardani, A. (1992) Analysis of Regional Growth and Disparity: The Impact of the Inpres Project on Indonesian Development. PhD thesis. University of Pennsylvania: USA.

Arrow, K.J. (1962) The Economic Implications of Learning by Doing, *Review of Economic Studies*, 29, pp. 155–173.

Asheim, B. (1996) Industrial district as learning region: a condition for prosperity, *European Planning Studies*, 4, pp. 379–400.

Asheim, B. and Coenen, L. (2006) Contextualising regional innovation systems in a globalising learning: on knowledge bases and institutional frameworks, *Journal of Technology Transfer*, 31 (1), pp. 163–173.

Bibliography

Asheim, B.T. and Cooke, P. (1999) Local Learning and Interactive Innovation Networks in a Global Economy, in Malecki, E. and Oinas, P. (eds) *Making Connections: Technological Learning and Regional Economic Change*. Aldershot: Ashgate, pp. 145–178.

Asheim, B.T. and Gertler, M. (2005) The geography of innovation: regional innovation systems, in Fagerbergh, J., Mowery, D. and Nelson, R. (eds) *The Oxford handbook of innovation*. Oxford: Oxford University Press.

Asheim, B.T., Gertler, M. and Vang, J. (2007) Face to face, buzz, and knowledge bases: Sociospatial implications for learning innovation and innovation policy, *Environment and Planning*, 25 (655–670), pp. 291–317.

Asheim, B.T. and Isaksen, A. (2002) Regional Innovation System: The Integration of Local Sticky and Global Ubiquitous Knowledge, *Journal of Technology Transfer*, 27, pp. 77–86.

Aswicahyono, H. and Anas, T. (2000) *Understanding the Pattern of Trade in the ASEAN Automotive Industry*. Jakarta: CSIS.

Aswicahyono, H., Anas, T. and Rizal, Y. (1999) *The Development of the Indonesian Automotive Industry*. Jakarta: CSIS.

Aswicahyono, H. and Feridhanusetyawan, T. (2004) *The Evolution and Upgrading of Indonesia's Industry*. Jakarta: CSIS.

Audretsch, D.B. and Feldman, M. (1996a) Innovative Clusters and the Industry Life-Cycle, *Review of Industrial Organization*, 11, pp. 253–273.

Audretsch, D.B. and Feldman, M. (1996b) R&D spillovers and the geography of innovation and production, *American Economic Review*, 86, pp. 630–640.

Audretsch, D. and Feldman, M.P. (2003) Knowledge Spillovers and the Geography of Innovation, in Henderson, J.V. and Thisse, J. (eds) *Handbook of Urban and Regional Economics*. Vol. 4 Amsterdam: North Holland Publishing.

Audretsch, D.B. and Lehmann, E.E. (2006) The role of clusters in knowledge creation and diffusion: an institutional perspective, in Asheim, B., Cooke, P. and Martin, R. (eds) *Clusters and Regional Development: Critical Reflections and Explorations*. Abingdon: Routledge, pp. 188–198.

Bathelt, H. (2006) Geographies of production: growth regimes in spatial perspective 3-towards a relational view of economic action and policy, *Progress in Human Geography*, 30, pp. 223–236.

Bathelt, H., Malmberg, A. and Maskell, P. (2004) Clusters and knowledge: Local buzz, global pipelines, and the process of knowledge creation, *Progress in Human Geography*, 28, pp. 31–56.

Baxter, J. and Eyles, J. (1997) Evaluating Qualitative Research in Social Geography Establishing 'Rigour' in Interview Analysis, *Transactions of the Institute of British Geographers*, 22, pp. 505–525.

Beck, U. (2005) *Power in the Global Age: A New Global Political Economy*. Cambridge: Cambridge University Press.

Bell, M. and Albu, M. (1999) Knowledge systems and technological dynamism in industrial clusters in developing countries, *World Development*, 27 (09).

Bell, M. and Pavitt, K. (1993) Technological Accumulation and Industrial Growth: Contrast between Developed and Developing Countries, *Industrial and Corporate Change*, 2, pp. 157–210.

Belussi, F. (2006) In search of a useful theory of spatial clustering: agglomeration versus active clustering, in Asheim, B., Cooke, P. and Martin, R. (eds) *Cluster and Regional Development: Critical Reflection and Explorations*. Abingdon: Routledge, pp. 69–89.

Benneworth, P. and Henry, N. (2003) Where is the value added in the cluster approach? Hermeneutic theorising, economic geography and cluster as a multiperspectical approach, *Urban Studies*, 41 (5–6), pp. 1011–1023.

Benneworth, P.S. and Charles, D.R. (2001) Bridging cluster theory and practice: Learning from the cluster policy cycle, in Bergman, E.M., Hertog, P.D., Charles, D.R. and Remoe, S. (eds) *Innovative clusters: drivers of national innovation systems*. Paris: OECD.

Best, M. (1990) *The New Competition: Institutions of Industrial Restructuring*, Cambridge: Polity Press.

Birkinshaw, J. (1996) How Multinational subsidiary mandates are gained and lost, *Journal of International Business*, 27, pp. 467–496.

Bloomberg, L.D. and Volpe, M. (2008) *Completing your qualitative dissertation: a roadmap from beginning to end*. Los Angeles: Sage.

Boekholt, P. (1994) *Methodology to Identify Clusters of Firms and Their Needs*. Luxemburg: paper for Sprint-RITTS workshop.

Boekholt, P. and Thruriaux, B. (1998) *Public Policies to Facilitate Clusters, Background, Rationale, and Policy Practices in International Perspective*. Paris: OECD.

Borrus, M. (1992) *Reorganizing Asia: Japan's New Development Trajectory and Regional Division of Labour*.

Boschma, R. and Wenting, R. (2006) *The Spatial Evolution of the British Automobile Industry*. Utrecht: Utrecht University.

Boschma, R.A. (2005) Proximity and Innovation: A Critical Assessment, *Regional Studies*, 39 (1), pp. 61–74.

Boschma, R.A. and Wenting, R. (2004) *The spatial evolution of the British automobile industry*. Utrecht: Urban and Regional Research Centre.

BPS (2004) *Pertumbuhan Ekonomi 2004*. Jakarta: BPS: Badan Busat Statistik.

BPS (2005) *Pertumbuhan Ekonomi 2005*. Jakarta: BPS: Badan Pusat Statistik.

Breschi, S. and Lissoni, F. (2001) Knowledge Spillovers and Local Innovation Systems: A Critical Survey, *Industrial and Corporate Change*, 10, pp. 975–1005.

Breschi, S. and Malerba, F. (2005) *Clusters, Networks, and Innovation*. Oxford: Oxford University Press.

Bresnahan, T., Gambardella, A. and Saxenian, A. (2001) Old Economy Inputs for New Economy Outcomes: Cluster Formation in the New Silicon Valleys, *Industrial and Corporate Change*, 10 (4), pp. 835–860.

Breziz, E.S. and Krugman, P. (1993) *Technology and the life-cycle of cities*. CBES.

Brown, P. and McNaughton, R. (2002) Global competitiveness and local networks: A review of the literature, in McNaughton, R. and Green, M. (eds) *Global competition and Local Networks*. London: Gower.

Bryson, J. and Henry, N. (2005) The global production system: from Fordism to post-Fordism, in Daniels, P., Bradshaw, M., Shaw, D. and Sideway, J. (eds) *An introduction to Human Geography: Issue for the 21st Century*. Harlow: Person, pp. 313–35.

Buckley, P.J. (2002) Is the international business research agenda running out of steam?, *Journal of International Business Studies*, 33 (2), pp. 365–73.

Buckley, P.J. and Chapman, M. (1996) Theory and method in international business research, *International Business Review*, 5 (3), pp. 233–45.

Burgess, R. and Venables, A. (2004) Towards a microeconomics of growth, in Bourguignon, F. and Pleskovic, B. (eds) *Annual Bank Conference in Development Economics: Accelerating Development*, Vol. 105–139 Oxford: Oxford University Press.

Bibliography

Burgess, R.G. (1991) Sponsor, gatekeepers, members, and friends, in Shaffir, W. B. and Stebbins, R. A.(eds) *Experiencing Fieldwork: an inside View of Qualitative Research*. London: Sage.

Busser, R. and Sadoi, Y. (2004) Introduction, in Sadoi, B. (ed.) *Production Networks in Asia and Europe: Skill formation and technology transfer in the automobile industry*. London: Routledge.

Caniëls, M.C.J. and Romijn, H.A. (2005) What drives innovativeness in industrial clusters? Transcending the debate, *Cambridge Journal of Economics*, 29 (4), pp. 497–515.

Cantner, U., Dressler, K. and Kruger, J.J. (2004) Firm survival in the German automobile industry, *Jenaer Schriften zur Wirtschaftswssenschaft*, 8, pp. 1–14.

Castells, M. (1996) The Rise of the Network Society. Cambridge: Blackwell.

Charles, D.R. and Li, F. (1993) *Lean production, supply chain management and new industrial dynamics: Logistic in the automobile industry*, Newcastle upon Tyne: CURDS.

Chen, M. (1996) *Managing international technology transfer*. London: International Thomson Business Press.

Coe, N., Dicken, P. and Hess, M. (2008) Global Production Network: Realizing the Potential, *Journal of Economic Geography*, 8, pp. 271–295.

Coe, N., Hess, M., Dicken, P., Yeung, H.W.C. and Henderson, J. (2004) Globalizing regional development: a global production networks perspective, *Transaction of the Institute of British Geographers*, 29, pp. 468–484.

Coe, N., Kelly, P.F. and Yeung, H.W.C. (2007) *Economic Geography: a contemporary introduction*. Oxford: Blackwell.

Cohen, W. and Levinthal, D. (1990) Absorptive Capacity: A new perspective on learning and innovation, *Administrative Science Quarterly*, 35, pp. 128–152.

Cooke, P. (1995) Regions, Clusters and Innovation Networks, in Cooke, P. (ed.) *The Rise of the Rustbelt*. London: UCL.

Cooke, P. (2001) From technopoles to Regional Innovation Systems: The evolution of localised technology development policy, *Canadian Journal of Regional Science*, XXIV (1), pp. 21–40.

Cooke, P. (2002) *Knowledge Economies*. London: Routledge.

Cooke, P. (2004) Regional Innovation Barriers and The Rise of Boundary Crossing Institution, in Wink, R. (ed.) *Academia–Business Links. European Policies and Lesson Learnt*. Basingstoke: Macmillan, pp. 223–242.

Cooke, P. (2006) *Between Implicit and Explicit Knowledge: Transnational Proximities and Innovation*, DIME.

Cooke, P., Heidenreich, M. and Braczyk, M. (2003) Regional Innovation Systems. London: Routledge.

Coombs, R., Narandren, P. and Richards, A. (1996) A literature-based innovation output indicator, *Research Policy*, 25, pp. 403–413.

Creswell, J.W. (1994) *Research Design: Qualitative and Quantitative Approaches*. Thousands Oaks, CA: Sage.

Dahl, M. and Pedersen, C.O.R. (2003) *Knowledge Flows through Informal Contacts in Industrial Cluster: Myths or Reality*. Copenhagen: DRUID.

Damuri, Y.R., Atje, R. and Gaduh, A.B. (2006) *Integration and Specialization in East Asia*. Jakarta: CSIS.

DeBresson, C. (1996) *Economics interdependence and innovative activity: an input–output analysis*. Cheltenham: Edward Elgar.

DeBresson, C. and Hu, X. (1997) *Techniques to identify innovative clusters: a method and 8 instruments*. Amsterdam: OECD.

DeBresson, C. and Hu, X. (1999) *Boosting Innovation: The Cluster Approach*. Paris: OECD.

Denscombe, M. (2000) *The Good Research Guide for Small-Scale Social Research Projects*. Buckingham: Biddles Ltd.

Dicken, P. (1988) The Changing Geography of Japanese Foreign Direct Investment in Manufacturing Industry: A Global Perspective, *Environment and Planning*, 20, pp. 633–653.

Dicken, P. (1992) *Global Shift: Industrial Change in a Turbulent World*, London: Paul Chapman.

Dicken, P. (2000) Places and flows: situating international investment, in Clark, G., Feldman, M. and Gertler, M. (eds) *The Oxford Handbook of Economic Geography*. Oxford: Oxford University Press, pp. 275–291.

Dicken, P. (2003) *Global Shift: Reshaping the Global Economic Map in the 21st Century*. London: Sage.

Dicken, P. (2005) *Tangled Webs: Transnational Production Networks and regional Integration*. Marburg: Phillips-University of Marburg.

Dicken, P. (2007) *Global shift: Mapping the changing contours of the World Economy*, London: Sage.

Dicken, P. and Henderson, J. (2003) *Making the Connections: Global Production Networks in Britain, East Asia and Eastern Europe*. Final report on ESRC Project R000238535.

Dicken, P. and Lyold, P.E. (1990) *Location in Space*. New York: HarperCollins.

Dicken, P. and Malmberg, A. (2001) Firms in Territories: A Relational Perspectives, *Economic Geography*, 77, pp. 345–363.

Dicken, P., Hudson, R. and Schamp, E.W. (1995) New Challenges to The Automobile Production Systems in Europe, in Hudson, R. and Schamp, E.W. (eds) *Towards a New Map of Automobile Manufacturing in Europe? New Production Concepts and Spatial Restructuring*. Vol. 1–20. Berlin.

Dohse, K., Jurgens, U. and Malsch, T. (1985) From Fordism to Toyotaism? The Social Organization of the labour Process in the Japanese Automobile Industry, *Politics and Society*, 14 (2).

Doner, R.F. (1991) *Driving a Bargain: Automobile Industrialization and Japanese Firms in Southeast Asia*. California: University of California Press.

Dowling, M. and Cheang, C.T. (2000) Shifting comparative advantage in Asia: New tests of the Flying Geese Model, *Journal of Asian Economics*, 11.

Drejer, I., Kristensen. F.S. and Laursen, K. (1999) *Studies of Clusters as a Basis for Industrial and Technology Policy and a Toolbox*. Paris: OECD.

Dunning, J.H. (1997) *Alliance Capitalism and Global Business*. London: Routledge.

Dunning, J.H. (2000a) Regions, Globalization, and the Knowledge Economy: The Issues Stated, in Dunning, J.H. (ed.) *Regions, Globalization, and The Knowledge-Based Economy*. Oxford: Oxford University Press, pp. 21–52.

Dunning, J.H. (2000b) Globalization and the Theory of MNE Activity, in Hood, N. and Young, S. (eds) *The Globalization of Multinational Enterprise Activity and Economic Development*. London: Macmillan Press Ltd, pp. 21–52.

Dupuy, C. and Torre, A. (2006) Local Cluster, Trust, Confidence, Proximity, in Christos, P., Sugden, R. and Wilson, J. (eds) *Clusters and Globalization: The Development of Urban and Regional Economies*. Cheltenham: Edward Elgar.

Dyer, J.H. and Nobeoka, K. (2000) Creating and managing a high-performance knowledge-sharing network: the Toyota case, *Strategic Management Journal*, 21, pp. 345–367.

Edquist, C. (2001) *The systems of innovation approach and innovation policy: an account of the state of the art*, DRUID Conference. Aalborg, 12–15 June 2001.

Edwards, R. (2002) FDI Strategic Issues, in Bora, B. (ed.) *Foreign Direct Investment: Research Issues*. London: Routledge, pp. 28–45.

Ernst, D. (1994) Carriers of Regionalization: The East Asian Production Networks of Japanese Electronic Firms, *Working paper on Berkeley Roundtable on The International Economy*, Vol. 73.

Ernst, D. (2000) Global Production Networks and the changing geography of innovation systems: Implication for Developing Countries, East-West Center, *Working Papers Economic Series*, 9.

Ernst, D. and Kim, L. (2002) Global Production Network, Knowledge Diffusion, and Local Capability Formation, *Research Policy*, 31, pp. 1417–1429.

Ernst, D. and Ravenhill, J. (2000) Convergence and Diversity: How Globalization Reshapes Asian Production, *East-West Working paper economies series*, 9.

Etzkowitz, H. (2006) The new visible hand: an assisted linear model of science and innovation policy, *Science and Public Policy*, 33 (5), pp. 310–320.

Etzkowitz, H., Dzisah, J., Ranga, M. and Zhou, C. (2007) The Triple Helix Model for Innovation: The University-industry-government interaction, *Asia Pacific Tech Monitor*, 24 (1), pp. 14–23.

Fagerbergh, J. (1993) *Technology and International Differences in Growth Rates*. Oslo: Norwegian Institute of International Affairs.

Fagerbergh, J. (1998) International Competitiveness, *Economic Journal*, 98, pp. 355–374.

FAIR (1989) *Interim Report of the Committee for Asia-Pacific Economic Research*. Singapore: Foundation for Advanced Information and Research.

Feldman, M., Aharonson, B. and Baum, J. (2005a) 'The Advancing Knowledge and The Knowledge Economy'. National Academics, Washington DC, 10–11 January.

Feldman, M.P. (2000) Location and Innovation: The New Economic Geography of Innovation, Spillovers, and Agglomeration, in Clark, G., Feldman, M. and Gertler, M. (eds) *Oxford Handbook of Economic Geography*. Vol. 373–394. Oxford: Oxford University Press.

Feldman, M.P. (2005) The Entrepreneurial Event Revisited: Firm Formation in a Regional Context. in Breschi, S. and Malerba, F. (eds) *Clusters, Networks, and Innovation*. Vol. 136–168. Oxford: Oxford University Press.

Feldman, M.P. and Francis, J. (2004) Home grown Solutions: Fostering Cluster Formation, *Economic Development Quarterly*, 18, pp. 127–137.

Feldman, M.P., Francis, J. and Bercovitz, J. (2005b) Creating a Cluster While Building a Firm: Entrepreneurs and the Formation of Industrial Clusters, *Regional Studies*, 39, pp. 129–141.

Florida, R. (1995) Towards the Learning Region, *Futures*, 27, pp. 527–536.

Florida, R., Jenkins, D. and Smith, D. (1998) The Japanese Transplants in North America: Production Organisation, Location, Research and Development, in Boyer, R., Charron, E., Jurgens, U. and Tilliday, S. (eds) *Between Imitation and Innovation: The Transfer and Hybridisation of Productive Models in the International Automobile Industry*. Oxford: Oxford University Press, pp. 189–215.

Freeman, C. (1987) *Technology and Economic Performance: Lessons from Japan*. London: Pinter.

Bibliography 191

Fujimoto, T. (2007) *Competing to be really really good: the behind the scenes drama of capability building competition in the automobile industry*. Tokyo: LTCB International Library Trust-International House of Japan.

Fujita, M. and Krugman, P. (1995) When is the Economy Monocentric? Von Thunen and Chamberlain United, *Regional Science and Urban Economics*, 25 (4), pp. 505–528.

Fujita, M. and Krugman, P. (2004) The New Economic Geography: Past, Present, and the Future, *Regional Science*, 83, pp. 139–164.

Fujita, M. and Thisse, J.F. (1996) *Economics of Agglomeration*. Centre for Economics Policy Research.

Fujita, M., Krugman, P. and Venables, A. (1999) *The spatial economy: cities, regions, and International trade*. Cambridge, MA: MIT Press.

GAIKINDO (2007) *Indonesian Automotive Statistic*. Available at: www.gaikindo.com (Accessed: 20 May 2007).

Gallaud, D. and Torre, A. (2004) Geographical Proximity and Circulation of Knowledge through Inter-firm Cooperation', in Wink, R. (ed.) *Academia–Business Link. European Policies and Lessons Learnt*. Basingstoke: Macmillan, pp. 137–158.

Gammeltoft, P. and Aminullah, E. (2004) The Indonesian innovation system at a crossroads, *1st ASIALICS International Conference: Innovation Systems & Clusters in Asia: Challenges & Regional Integration*. Bangkok, April 1–2 2004. The Indonesian innovation system at a crossroads.

Garnsey, E. and Longhi, C. (2004) High technology locations and globalization: converse-paths, common processes, *International Journal of Technology and Management*, 28, pp. 336–367.

Gereffi, G. (1994) The organisation of buyer-driven global commodity chains: how US retailers shape overseas production networks, in Gereffi, G. and Korzeniewicz, M., *Commodity chains and global capitalism*, Wesport: CT Praeger.

Gereffi, G. (2001) Beyond the producer-driven buyer-driven dichotomy. The evolution of Global Value Chains in the Internet Era, *IDS Bulletin*, 32 (3).

Gereffi, G., Humphrey, J. and Sturgeon, T. (2005) The governance of global value chains, *Review of International Political Economy*, 12, pp. 78–104.

Gerlach, M. (1992) *Alliance Capitalism: The Social Organization of Japanese Business*. Berkeley: University of California Press.

Gertler, M.S. (1995) Being there: Proximity, Organization and Culture in the Development and Adoption of Advanced Manufacturing Technologies, *Economic Geography*, 71 (1), pp. 1–26.

Gertler, M.S. (2003) Tacit Knowledge and The Economic Geography of Context, Or The Undefinable Tacitness of Being There, *Economic Geography*, 3 (1), pp. 75–99.

Gertler, M.S. and Wolfe, D.A. (2006) Spaces of knowledge flows: clusters in a global context, in Asheim, B., Cooke, P. and Martin, R. (eds) *Clusters and Regional Development: Critical Reflections and Explorations*. Abingdon: Routledge, pp. 218–235.

Ginzburg, A. and Simonazzi, A. (2004) Patterns of industrialization and the flying geese model: the case of electronics in East Asia, *Journal of Asian Economics*, 15, pp. 1051–1078.

Giuliani, E. (2006) The selective nature of knowledge networks in clusters: evidence from the wine industry, *Journal of Economic Geography*, 7, pp. 139–168.

Giuliani, E., Bell, M. and Pavitt, K. (2005) The micro determinants of meso-level learning and innovation: evidence from a Chilean wine cluster, *Research Policy*, 34, pp. 47–68.

Bibliography

Glaeser, E.L. (2000) The new economics of urban and regional growth, in Clark, G.L., Feldman, M.P. and Gertler, M.S. (eds) *Oxford Handbook of Economic Geography*. Vol. 83–98 Oxford: Oxford University Press.

Gordon, I.R. and McCann, P. (2000) Industrial Cluster: Complexes, Agglomeration, and or Social Network, *Urban Studies*, 37 (3), pp. 513–532.

Gordon, I.R. and McCann, P. (2005) Innovation, Agglomeration and Regional Development, *Journal of Economic Geography*, 5, pp. 523–543.

Gorgh, H. and Greenaway, D. (2002) *Do domestic firms really benefit from FDI?*. Nottingham: Centre for research on globalisation.

Goyal, S. and Joshi, S. (2003) Networks of Collaboration in Oligopoly, *Games and Economic Behaviour*, 43, pp. 57–85.

Grabher, G. (1993) *The Embedded Firm: On the Socio-economics of Industrial Networks*. London: Routledge.

Gunawan, R. (2002) *The Short Analysis of Motorcycle's Market and Industries in Indonesian for the year 2000 and 2001*. Available at: www.aisi.or.id/wnew.html (Accessed: 22 July 2008).

Han, M.H. (1994) *Japanese Multinationals in the Changing Context of Regional Policy*. Hants: Ashgate Publishing Limited.

Hanson, G.H. (2001) Should countries promote foreign direct investment?, *G-24 Discussion Paper No. 9*. New York: United Nations.

Harada, T. (2003) Three Steps in Knowledge Communication: the emergence of knowledge transformers, *Research Policy*, 32, pp. 1737–1751.

Harrison, B. (1994) *Lean and Mean: The changing landscape of corporate power in the age of flexibility*. New York: Basic Books.

Harrison, B., Kelley, M. and Grant, J. (1996) Innovative firm behaviour and local milieu: Exploring the intersection of agglomeration, firm effects, industrial organization, and technological change, *Economic Geography*, 72, pp. 233–258.

Hatch, W. and Yamamura, K. (1996) *Asia in Japan's Embrace: Building a Regional Production Alliance*. Cambridge: Cambridge University Press.

Held, J.R. (1996) Clusters as an Economic Development Tool: Beyond the pitfalls, *Economic Development Quarterly*, 10 (3), pp. 249–261.

Henderson, J. (2002) Global Production Network and The Analysis of Economic Development, *Review of International Political Economy*, 9, pp. 436–464.

Henderson, J.V., Lee, T. and Lee, Y.J. (2001) Scale Externalities in Korea, *Journal of Urban Economics*, 49, pp. 479–504.

Hess, M. and Yeung, H.W.C. (2006) Whether global production networks in economic geography? Past, present and future, *Environment and Planning*, 38, pp. 1193–1204.

Hideki, Y. (1988) *The Global Economy of The Japanese Firms*. Tokyo: Toyo Keizai Shinposa.

Hieneman, B.D., Johnson, C., Pamani, A. and Park, H.J. (1985) Technology Transfer from Japan to Southeast Asia, in Samli, A.C. (ed.) *Technology Transfer: Geographic, Economic, Cultural and Technical Dimension*. Westport, Conn: Quorum Books, pp. 143–153.

Hill, H. (1997) *Indonesia's Industrial Transformation*. Singapore: Institute of Southeast Asian Studies.

Hill, H. (2001) Small and Medium Enterprises, *Indonesia Asian Survey*, 41 (2), pp. 248–270.

Hino, S. (2005) *Inside the Mind of Toyota: Management Principles for Enduring Growth*. University Park, IL: Productivity Press.

HKI (2006) *Indonesia Industrial Estate Directory: A Guide for Investors*. Jakarta: HKI: Indonesian Industrial Estate Association.

Hoen, A.R. (2002) Identifying Linkages with a Cluster-based Methodology, *Economic Systems Research*, 14 (2), pp. 131–146.

Holm, U., Johanson, J. and Thilenieus, P. (1995) Headquarter knowledge of subsidiary contexts in the multinational corporation, *International Studies of Management and Organization*, 25 (1–2), pp. 97–119.

Honda (1991) *Fact and Figure*. Honda Motor: Tokyo.

Honda (2004) *Honda Motor Co. Ltd: World Motorcycle: Facts and Figures*. Tokyo.

Honda (2007a) *Astra Honda Indonesia: Annual Report*. Jakarta: Astra Honda Indonesia.

Honda (2007b) *Honda in Indonesia*. Available at: www.world.honda.com/Indonesia/ (Accessed: 29 May 2007).

Hoover, E. (1937) *Location Theory and the shoe and leather industries*. Cambridge: Harvard University Press.

Hotelling, H. (1929) Stability in competition, *Economic Journal*, 39, pp. 41–57.

Humphrey, J., Lecler, Y. and Salerno, M.S. (2000) Introduction, in Humphrey, J., Lecler, Y. and Salerno, M.S. (eds) *Global Strategies and Local Realities. The Auto Industry in Emerging Markets*. Hampshire: Houndsmills, Basingstoke, pp. 1–15.

Humphrey, J. and Schmitz, H. (2000) Governance and Upgrading: Linking Industrial Cluster and Global Value Chain Research, *IDS Working Paper*, 120.

Iammarino, S. and McCann, P. (2006) The structure and evolution of industrial cluster: transactions, technology and knowledge spillovers, *Research Policy*, 35, pp. 1018–1036.

Ichiro, S. (1991) Localization Policy for Automobile Production, in Doner, R. (ed.) *Driving a Bargain: Automobile Industrialization and Japanese Firms in Southeast Asia*. Berkeley: University of California Press, p. 80.

IMF (2007) *World Economic Outlook*. Washington DC: International Monetary Fund.

Jababeka (2005) *The Jababeka industrial cluster*. Available at: www.jababeka.com (Accessed: 23 November 2005).

Jacobs, D. and de Man, A.-P. (1996) Cluster, industrial policy and firm strategy: a menu approach, *Technology Analysis and Strategic Management*, 8, pp. 425–437.

Jacobson, D. and Andrèosso-O'Callaghan, B. (1996) *Industrial economics and organization: A European perspective*, McGraw-Hill: London

Jaffe, A., Trajtnberg, M. and Henderson, R. (1993) Geographic localization of knowledge spillovers as evidenced by patents citations, *Quarterly Journal of Economics*, 108, pp. 577–598.

JAMAI (2008) *The Motor industry of Japan*. Tokyo: Japan Automobile Manufacturers Association Inc.

Jarillo, J.C. (1993) *Strategic Networks: Creating the Borderless organisations*. Oxford: Butterworth-Heinemann.

Jonash, D.T. and Womack, J.P. (1985) Developing Countries and the Future of The Automobile Industry, *World Development*, 13 (3), pp. 393–407.

Kadoto, Y. (1985) *Toyota Shisutemu (Toyota System)*. Tokyo: Kodansha.

Kayoko, K. (1991) Developments and the Status of Direct Investment in Asia, *The Second Asia-Pacific Conference Tokyo*, 9–10 May 1991.

Kenney, M. and Florida, R. (1993) *Beyond Mass Production: The Japanese System and its Transfer to the US*. New York: Oxford University Press.

Kiba, T. and Kodama, F. (1991) *Measurement and Analysis of the Progress of International Technology Transfer: Case Study of Direct Investment in East Asian Countries by Japanese Companies*. Japan: National Institute of Science and Technology Policy, Science, and Technology Agency of Japan.

Bibliography

Kim S.-R. and Von Tunzelmann, N. (1998) Aligning internal and external networks: Taiwan specialisation in IT, *SEWPS*, 17.

Klepper, S. (2001) *The Evolution of the U.S. Automobile Industry and Detroit as its Capital*. Mimeo: Carnegie Mellon University.

Klepper, S. (2002) The capabilities of new firms and the evolution of the US automobile industry, *Industrial and Corporate Change*, 11 (4), pp. 645–66.

Klepper, S. (2004) *Agglomeration through spinoffs: How Detroit became the capital of the US automobile industry*. Carnegie Mellon University.

Kodama, F. and Kiba, T. (1994) *The Emerging Trajectory of International Technology*. Stanford: Asia-Pacific Research Center-Stanford University.

Kogut, B. (2000) The Network as Knowledge: Generative Rules and The Emergence of Structure, *Strategic Management Journal*, 21, pp. 405–425.

Krugman, P. (1991a) *Geography and Trade*. Cambridge: MIT Press.

Krugman, P. (1991b) Increasing returns and economic geography, *Journal of Political Economy*, 99, pp. 483–499.

Krugman, P. and Venables, A. (1995) Globalization and the Inequality of nations, *Quarterly Journal of Economics*, 110, pp. 857–880.

Kuncoro, M. (1996) Regional Development in Indonesia: Some Notes towards 21st Century. *UNISIA*, 31.

Kuncoro, M. (2002) *Analisis Spasial dan Regional: Studi Aglomerasi dan Kluster Industri Indonesia*. Yogyakarta: UPP AMP YKPN.

Kuznetsov, Y. and Sabel, C. (2007) *Towards a new open economy industrial policy: Sustaining growth without picking winners*. Washington DC: World Bank Institutes.

Lagendijk, A. and Charles, D. (1997) OECD *workshop on cluster analysis and cluster based policy*. Amsterdam, 10–11 October.

Lagendijk, A. and Charles, D. (1999) *Clustering as a new growth strategy for regional economies?* Paris: OECD.

Lall, S.V., Funderbugh, R. and Yepes, T. (2004) *Location, Concentration, and Performance of Economic Activity in Brazil*. Policy Research Working Paper. World Bank.

Larsson, A. (1999) Proximity Matters? Geographical aspects of changing strategies in automotive subcontracting relationships: the case of domestic suppliers to Volvo Torslanda assembly plant. Thesis. University of Goteborg.

Lazonick, W. (1992) *Industry Clusters versus Global Webs*. New York: Department of Economics. Columbia University.

Leclerc, E. and Meyer, J.B. (2007) Knowledge diasporas for development: A shrinking space for skepticism, *Asian Population Studies*, 3, pp. 153–168.

Leonard, D.A. (1984) Swedish Entrepreneurs in Manufacturing and their Sources of Information. *Research Policy*, 13 (April), pp. 101–114.

Liker, J. (2004a) *The 14 Principles Of The Toyota Way: An Executive Summary of the Culture Behind TPS*. McGraw-Hill.

Liker, J. (2004b) *The Toyota Way: 14 Management Principles from the World's Greatest Manufacturer*. McGraw-Hill.

Liker, J. and Meier, D. (2005) *The Toyota Way Field book: A Practical Guide for Implementing Toyota's 4Ps*. McGraw-Hill.

Lindsey, C. (1985) Conference in ASEAN–US Economic Relations. Singapore, 22–24 April 1985.

Liu, W. and Dicken, P. (2006) Transnational corporations and obligated embeddedness: foreign direct investment in China's automobile industry, *Environment and Planning*, 38, pp. 1229–1247.

Lucas, R. Jr. (1988) On the mechanics of economic development, *Journal of Monetary Economics*, 22(1), pp. 3–42.

Lundan, S. (2002) *Network Knowledge in International Business*. Cheltenham: Edward Elgar.

Lundvall, B.A. (ed.) (1998) *Innovation as an interactive process: from user-producer interaction to the national system of innovation*. London: Pinter Publisher.

Lundvall, B.A. and Christensen, J.L. (1999) Extending and deepening the analysis of innovation systems-with empirical illustration from the DISCO-project. *DRUID Working Paper*.

Maher, M. (2001) Benefit *and costs of foreign direct investment: A survey*. Paris: OECD.

Mai, K.-I. (1991) *Globalization and Cross-Border Network of Japanese Firms*. Stanford-Japan Centre Research.

Mair, A. (1994) *Honda's Global Local Corporation*. New York: St. Martin's Press, Inc.

Malecki, E. and Nijkamp, P. (1988) Technology and regional development: some thoughts on policy, *Environment and Planning: Government and Policy*, 6, pp. 383–399.

Malerba, F. and Orsenigo, L. (ed.) (1990) *Technological regimes and patterns of innovation: a theoretical and empirical investigation of the Italian case*. Cambridge: Cambridge University Press.

Malerba, F. (2002) Sectoral Systems of Innovation and Production, *Research Policy*, 31 (2), pp. 247–267.

Malerba, F. (2005) Sectoral System of Innovation: a framework for linking innovation to knowledge base, structure, and dynamic sectors, *Economics of innovation and new Technology*, 14 (1), pp. 63–82.

Malmberg, A. (2003) Beyond the cluster-local mileux and global economic connections, in Peck, J. and Yeung, H.W.C. (eds) *Remaking the global economy*. London: Sage, pp. 145–159.

Malmberg, A. and Maskell, P. (1999a) Localized Learning and Regional Development, *European Urban and Regional Studies*, 6 (1), pp. 5–8.

Malmberg, A. and Maskell, P. (1999b) The Competitiveness of Firms and Regions. Ubiquitification and the Importance of Localized Learning, *European Urban and Regional Studies*, 6 (1), pp. 9–25.

Malmberg, A. and Maskell, P. (2004) The elusive concept of localization economies: towards a knowledge-based theory of spatial clustering, in Grabher, G. and Powell, W.W. (eds) *Networks*. Edward Elgar Series Critical Studies in Economic Institutions, Cheltenham: Edward Elgar.

Malmberg, A. and Maskell, P. (2006) Localized Learning Revisited, *Growth and Change*, 37 (1), pp. 1–18.

Malmberg, A. and Power, D. (2005) How do firms in clusters create knowledge?, *Industry and Innovation*, 12, pp. 409–431.

Malmberg, A. and Power, D. (2006) True cluster: a severe case of conceptual headache, in Asheim, B., Cooke, P. and Martin, R. (eds) *Clusters and Regional Development: Critical Reflections and Explorations*. Abingdon: Routledge, pp. 429–449.

Malmberg, A. and Sölvell, O. (2002) Does Foreign Ownership Matter? Subsidiary Impact on Local Clusters, in Havila, V., Forsgren, M. and Hakansson, H. (eds) *Critical Perspectives on Internationalization*. Amsterdam: Pergamon, pp. 59–78.

Malmberg, A., Sölvell, O. and Zander, I. (1996) Spatial Clustering, Local Accumulation of Knowledge and Firm Competitiveness, *Geografiska Annaler*, 78b, pp. 1992–2.

Markusen, A. (1996) Sticky Places in Slippery Space: A Typology of Industrial Districts, *Economic Geography*, 72 (3), pp. 293–313.

Bibliography

Markusen, A. (1999) Fuzzy concepts, scanty evidence, policy distance: the case for rigour and policy relevance in critical regional studies, *Regional Studies*, 9, pp. 869–894.

Marschan-Piekkari, R. and Welch, C. (2004) *Handbook of Qualitative Research: Methods for International Business*. Cheltenham: Edward Elgar.

Marshall, A. (1890) *Principles of Economics*. London: Macmillan.

Martin, R. (1999) The new 'geographical turn' in economies: Some critical reflections, *Cambridge Journal of Economics*, 23 (1), pp. 65.

Martin, R. and Sunley, P. (2003) Deconstructing Clusters: Chaotic Concept or Policy Panacea?, *Economic Geography*, 3 (1), pp. 5–35.

Martin, R. and Sunley, P. (2006) Path Dependence and Regional Economic Evolution, *Journal of Economic Geography*, 6 (4), pp. 395–437.

Maskell, P. (2001) Towards a knowledge-based theory of the geographical cluster, *Industrial and Corporate Change*, 10 (4), pp. 3–22.

Maskell, P., Malmberg, A. and Bathelt, H. (2006) Building Global Knowledge Pipelines: The Role of Temporary Cluster, *European Planning Studies*, 14, pp. 997–1013.

Mason, J. (2002) *Qualitative Researching*. Thousand Oaks, CA: Sage.

Mathew, J. and Cho, D.S. (2000) *Tiger Technology*. Cambridge: Cambridge University Press.

Maxwell, J.A. (1996) *Qualitative Research Design: An Interactive Approach*. Thousand Oaks, CA: Sage.

McCann, P. (ed.) (1998) *The Economic of Industrial Location: A logistic-cost Approach*. Heidelberg.

Meeuwsen, W. and Dumont, M. (1997) *Some result on the graph-theoretical identification of micro-clusters in the Belgian National Innovation System*. Amsterdam: OECD.

Metcalfe, S. (ed.) (1995) *The economic foundations of technology policy*. Oxford: Oxford University Press.

Metcalfe, S. and Ramlogan, R. (2006) Innovation Systems and The Competitive Process, *Developing Economies, Regulation, competition and income distribution: Latin American experiences*. Brazil, November 18–21, 2005.

Meyer-Stamer, J., Maggi, C. and Seibel, S. (2001) *Improving upon nature. Creating competitive advantage in ceramic tile clusters in Italy, Spain, and Brazil*. INEF.

Mikler, J.J. (2007) Varieties of capitalism and the auto industry's environmental initiatives: national institutional explanations for firms' motivations, *Business and Politics*, 9.

Mito, S. (1990) *The Honda Book of Management: A Leadership Philosophy for High Industrial Success*. London: The Athlone Press Ltd.

Miyakawa, Y. (1991) The transformation of the Japanese Motor Vehicle Industry and Its Role in the World: Industrial Restructuring and Technical Evolution, in Law, M.C. (ed.), *Restructuring the Global Automobile Industry*. London: Routledge.

Morgan, K. (1997a) The Learning Region: Institutions, Innovation and Regional renewal, *Regional Studies*, 31, pp. 491–503.

Morgan, K. (1997b) *New Approaches in Technology Policy – The Finnish Example*. Vienna: OECD.

Morgan, K. and Nauwelaers, C. (eds) (1999) *Regional innovation strategies: the challenge for less-favoured regions*. London: The stationary office with the Regional Studies Association.

Morgan, K. and Nauwelaers, C. (2003) *Regional Innovation Strategies: The Challenge for Less-Favoured Regions*. London: Routledge.

Bibliography 197

Morrison, A. and Rabelloti, R. (2005) *Inside the black box of industrial atmosphere: Knowledge and information networks in an Italian Wine Local System.* Novara: Università Del Piomonte Orientale'.

MOSR (2002) *The industrial and research policy in Indonesia.* Jakarta: Ministry of Science and Research, Republic of Indonesia.

MOTIRI (2005) *The Automotive Industry and its progress.* Jakarta: Ministry of Trade and Industry, Republic of Indonesia.

Moulaert, F. and Swyngedouw, E. (1991) Regional development and the geography of the flexible production system: theoretical arguments and empirical evidence, in Hilpert, U. (ed.) *Regional innovation and decentralization: Hi-tech industry and government policy.* London: Routledge.

Murphy, K.M., Shleifer, A. and Vishny, R. (1989) Industrialization and the Big Push, *Journal of Political Economy*, 97, pp. 1003–1026.

Myers, M.B. and Rosenbloom, R.S. (1996) Rethinking the role of research, *Research Technology Management*, 39(3), pp. 14–18.

Narula, R. (2003) *Globalisation and Technology: Interdependence, Innovation Systems and Industrial Policy.* Cambridge: Polity Press.

Narula, R. and Dunning, J. (2002) Industrial Development, Globalization and Multinational Enterprises: New Realities for Developing Countries, *Oxford Development Studies*, 28 (2), pp. 141–167.

Narula, R. and Zanfei, A. (2005) Globalization of Innovation: The Role of Multinational Enterprises, in Fagerbergh, J., Mowery, D.C. and Nelson, R.R. (eds) *The Oxford Handbook of Innovation.* Oxford: Oxford University, pp. 318–347.

Nicholas, S. and Maitland, E. (2002) International Business Research: Steady States, Dynamics and Globalisation, in B. Bora (ed.) *Foreign Direct Investment: Research Issues.* London: Routledge, pp. 7–27.

Nijkamp, P. (1999) De Revival van den Regio, in Graaff, W.D. and Boekema, F. (eds) *De Regio Centraal.Assen*: *Van Gorkum & Comp.* Nikei weekly, 20 June 1994, pp. 1–23.

Nomura, M. (1993) *Toyotizumu (Toyotaism).* Tokyo: Mineruba.

Nonaka, I. and Takeuchi, H. (1995) *The Knowledge Creating Company: How Japanese Companies Create the Dynamics of Innovation.* Oxford: Oxford University Press.

Nonaka, I., Toyama, R. and Nagata, R. (2000) The Firms as the Knowledge-Creating Entity: A new Perspective on the Theory of The Firm, *Industrial and Corporate Change*, 9 (1–20).

Nooteboom, B. (2003) Problems and solution in knowledge transfer, in T, B. (ed.), *The influence of cooperation: networks and innovation on regional innovation systems.* Cheltenham: Edward Elgar.

Nooteboom, B. (2004) *Inter-firm collaboration, Learning networks: An integrated approach.* London: Routledge.

Nooteboom, B. (2006) Innovation, learning and cluster dynamics, in Asheim, B., Cooke, P. and Martin, R. (eds) *Clusters and Regional Development: Critical Reflections and Explorations.* Abingdon: Routledge, pp. 137–163.

Odaka, K., Ono, K. and Adachi, F. (1988) *The Automobile Industry in Japan: A Study of Ancillary Firm Development.* Tokyo: Kokusabunken Insatsusha.

OECD (1999a) *Boosting Innovation: The cluster approach.* Paris: OECD.

OECD (1999b) *Managing Innovation Systems.* Paris: OECD.

OECD (2008) *Indonesia Economic Assessment.* Paris: OECD.

Okada, Y. (1983) The Dilemma of Indonesian Dependency on FDI, *Development and Change*, 14.

Osland, J. and Osland, A. (2001) International qualitative research: an effective way to generate and verify cross-cultural theories, in Toyne, B., Martinez, Z.L. and Menger, R.A. (eds) *International Business Scholarship*. Vol. 198–214 Westport, CT: Quorum.

Ozawa, T. (1981) *Transfer of Technology from Japan to Developing Countries*. New York: UNITAR.

Ozawa, T. (2005) *Institutions, Industrial Upgrading, and Economic Performance in Japan – The 'Flying-Geese Paradigm of Catch-up Growth*. Northampton, Massachusetts: Edward Elgar Publishing.

Parrilli, M.D. (2006) A tripartite innovation framework for small firms in developing regions: key issues for analysis and policy, in Bianchi, P. and Labory, S. (eds) *Handbook on industrial development policy*. Cheltenham: Edward Elgar.

Patel, P. and Pavitt, K. (eds) (1998) *Uneven (and Divergent) Technological Accumulation among Advanced Countries: Evidence and a Framework of Explanation*. Oxford: Oxford University Press.

Pavitt, K. (1987a) The objectives of technology policy, Science and Public Policy, 14, pp. 182–188.

Pavitt, K. (1987b) *On the nature of technology*. Mimeo: University of Sussex-Science Policy Research Unit.

Pavitt, K. (2005) Innovation processes, in Fagerbergh, J., Mowery, D.C. and Nelson, R.R. (eds) *The Oxford Handbook of Innovation*. Oxford: Oxford University Press, pp. 86–114.

Pedersen, T., Petersen, B. and Sharma, D.D. (2003) *Knowledge Transfer Performance of Multinational Companies*, Management International Review, 43, pp. 69–90.

Pelegrín, A. and Bolancé, C. (2008) Regional Foreign Direct Investment in Manufacturing Do Agglomeration Economies Matter?, *Regional Studies*, 42 (4), pp. 505–522.

Perdana, A. and Friawan, D. (2007) *Economic Crisis, Institutional Changes, and the Effectiveness of Government: The Case of Indonesia*. Jakarta: CSIS.

Phelps, N.A. (2008) 'Cluster or Capture? Manufacturing Foreign direct Investment, External Economies and Agglomeration, *Regional Studies*, 42 (4), pp. 457–473.

Pinch, S. and Henry, N. (1999) Paul Krugman's geographical economics, industrial clustering, and the British motor sport industry, *Regional Studies*, 33, pp. 815–27.

Piore, M.J. and Sabel, C.F. (1994) *The Second Industrial Divide, Possibilities for Prosperity*. New York: Basic Books.

Poon, J.P.H., Hsu, J.Y. and Jeongwook, S. (2006) The Geography of learning and knowledge acquisition among Asian late comers, *Journal of Economic Geography*, 6, pp. 541–559.

Porter, M.E. (1998a) Cluster and the New Economic of Competition, *Harvard Business Review*, November–December, pp. 78–90.

Porter, M.E. (ed.) (1998b) *Cluster and competition: New agendas for companies, governments, and institutions*. Boston: Harvard Business School Press.

Porter, M.E. (1998c) *The microeconomics foundations of economic development'*, World Economic Forum. Geneva, The microeconomics foundations of economic development: pp. 38–63.

Porter, M.E. (1990) *The Competitive Advantage of Nations*. Basingstoke: Macmillan.

Porter, M.E. (2000) Location, Competition and Economics Development: Local clusters in a global economy, *Economic Development Quarterly*, 14 (1), pp. 15–34.

Powell, W.W. and Grodal, S. (2005) Networks of innovators, in Fagerbergh, J., Mowery, D.C. and Nelson, R.R. (eds) *The Oxford Handbook of Innovation*. Oxford: Oxford University Press, pp. 56–85.

Power, D. and Lundmark, M. (2004) Working through Knowledge Pools: labour market dynamics, the transference of knowledge and ideas and industrial clusters, *Urban studies*, 41, pp. 1025–1044.

Prahalad, C.K. and Ramaswamy, V. (2004) *The Future of Competition*. Boston, MA: Harvard Business School Press.

Pries, L. and Schweer, O. (2004) The Product Development Process as a Measuring Tool for Company Internationalization, *International Journal of Automotive Technology and Management*.

Rasiah, R. (2004) *Foreign Firms, Technological Capabilities and Economic Performance: Evidence from Africa, Asia, and Latin America*. Cheltenham: Edward Elgar.

Rasiah, R. (2005) Foreign Ownership, Technological Intensity and Export Incidence: A study of Auto Parts, Electronics and Garment Firms in Indonesia, *International Journal of High Technology and Globalization*, 1, pp. 361–380.

Rasiah, R., Sadoi, Y. and Busser, R. (2008) Special Issue: Multinationals, Technology and Localization in the Automotive Industry in Asia, *Asia Pacific Business Review*, 14 (1), pp. 1–12.

Rikard, E. and Urban, L. (2008) Localized mobility clusters: impacts of labour market externalities on firm performance, *Journal of Economic Geography*, 2, pp. 1–21.

Rodrik, D. (1991) Policy Uncertainty and Private Investment in Developing Countries, *Journal of Development Economics*, Vol. 36 (November).

Rodriguez-Clare, A. (1996) Multinationals, Linkages and Economic Development, *American Economic Review*, 86 (4), pp. 852–73.

Roelandt, T.J.A., Den Hertog, P., Van Sinderen, J. and Van den Hove, N. (1999) *Boosting Innovation: The Cluster Approach*. Paris: OECD.

Romer, P.M. (1986) Increasing Returns and Long-run Growth, *Journal of Political Economy*, 94 (5), pp. 1002–1037.

Romjin, H. (2002) Acquisition of technological capability in small firms in LDCs, in Van Dijk, M. and Sandee, H. (eds) *Innovation and small firms in the Third World*. London: Edward Elgar.

Rosenthal, S. and Strange, W.C. (2003) Evidence on the nature and sources of agglomeration economies, in Henderson, V. and Thisse, J. (eds) *Handbook of Urban and Regional Economics*. Vol. 4.

Rugman, A.M. and D'Cruz, J.R. (2002) *Multinationals as flagship firms*. Oxford: Oxford University Press.

Rutherford, T. (2000) Re-embedding Japanese Investment and the Restructuring Buyer-Supplier Relations in the Canadian Automotive Components Industry During the 1990s, *Regional Studies*, 34 (8), pp. 739–751.

Ryen, A. (2002) Cross cultural interviewing, in Gubrium, J.F. and Holestain, J.A. (eds) *Handbook of Interview Research: Context and Method*. Thousand Oaks, CA: Sage, pp. 335–353.

Sakiya, T. (1982) *Honda Motor: The Men, The Management, The Machines*. Tokyo: Kondansha International Ltd.

Samuels, R.J. (1987) *The Business of The Japanese State: Energy Markets in Comparative and Historical Perspective*. Ithaca: Cornell University Press.

Sandee, H. (1994) The Impact of Technological Change on Interfirm Linkages. A Case Study of Clustered Rural Small-Scale Roof Tile Enterprises in Central Java, in Pedersen, P.O., Sverrison, A. and van Dijk, M.P. (eds) *Flexible Specialization. The Dynamics of Small-Scale Industries in the South*. London: Intermediate Technology Publications, pp. 21–40.

Bibliography

Sandee, H. and ter Wingel, J. (2002) SME Cluster Development Strategies in Indonesia: What Can We Learn from Successful Clusters? Paper presented at the JICA Workshop on Strengthening Capacity of SME Clusters in Indonesia, Jakarta, 5–6 March.

Sato, Y. (2001) *Structure, Features and Determinants of Vertical Inter-firm Linkages in Indonesia*. Thesis. University of Indonesia.

Saxenian, A. (1996) *Regional Advantage: Culture and Competition in Silicon Valley and Route 128*. Harvard: Harvard University Press.

Saxenian, A. (2002) Transnational Communities and The Evolution of Global Production Networks: The Case of Taiwan, China, and India, *Industry and Innovation*, 9, pp. 183–202.

Saxenian, A. (2005) District and Diaspora: Local and Global Knowledge in Networked Economy, *The Advancing Knowledge and the Knowledge Economy*. Washington DC.

Saxenian, A. (2008) Lecture in Economic Geography-Venture Capital in the Periphery: The New Argonauts, Global Search, and Local Institution Building, *Economic Geography*, 84 (4), pp. 379–394.

Schmitz, H. (1982) Growth constraints on small-scale manufacturing in developing countries: a critical review, *World Development*, 10 (6).

Schmitz, H. (1995) Collective Efficiency: growth path for small-scale Industry, *Journal of Development Studies*, 31 (4).

Schmitz, H. (2000) Does Local Co-Operation Matter? Evidence from Industrial Clusters in South Asia and Latin America, *Oxford Development Studies*, 28 (3), pp. 323–336.

Schnepp, O., Glinow, V., Ann, M. and Arvind, B. (1990) *United States–China Technology Transfer*. Englewood Cliffs, NJ: Prentice Hall.

Schumpeter, J. (1934) *The Theory of Economic Development*. Cambridge: Harvard University Press.

Scott, A.J. (1982) Locational patterns and dynamics of industrial activity in the modern metropolis, *Urban Studies*, 19, pp. 111–141.

Scott, A.J. and Storper, M. (1987) High Technology Industry and Regional Development: A Theoretical Critique and Reconstruction, *International Social Science Journal*, 112, pp. 215–232.

Scott, J. (1991) *Social Network Analysis: A handbook*. London: Sage.

Seale, C., Gobo, G., Gubrium, J.F. and Silverman, D. (2004) *Qualitative Research Practice*. London: Sage.

Sekiguchi, S. (1983) *ASEAN-Japan Relations: Investment*. Singapore: Institute of Southeast Asian Studies.

Shimizu, K. (1999) *Le Toyotisme*. Paris: La Dèuverte.

Silverman, D. (2004) *Qualitative Research: Theory, Method and Practice*. Thousand Oaks, CA: Sage.

Silverman, D. (2005) *Doing Qualitative Research*. Thousand Oaks, CA: Sage.

Smitka, M.J. (1991) *Competitive Ties: Subcontracting in the Japanese Automotive Industry*. New York: Columbia University Press.

Soesastro, H. (2003) *Economic Recovery and Reform in Indonesia*. Jakarta: CSIS.

Soesastro, M.H. (1989) *Japan–Southeast Asia*. Kuala Lumpur, 24–27 November.

Solimano, A. (2008) *The international mobility of talent: Types, causes, and development impact*. Oxford: Oxford University Press.

Sonis, M., Hewings, G.J.D., Guo, J. and Hulu, E. (1997) Interpreting Spatial Economic Structure: Feedback Loops in the Indonesian Interregional Economic, *Regional Science and Urban Economics*, 27, pp. 325–342.

Stake, R. (2000) Case Studies is a good account of the conventional qualitative methods position on generalizability, in Denzin, N. and Lincoln, Y. (eds) *Handbook of Qualitative Research*. Thousand Oaks, CA: Sage.
Steiner, M. (1987) Contrasts in regional potentials: some aspects of regional economic development, *Regional Science Association*, 61, pp. 79–92.
Steiner, M. (ed.) (1998) *Clusters and Regional Spatialisation*. London: Pion Limited.
Steiner, M. (2006) Do clusters think? an institutional perspective on knowledge creation and diffusion in clusters, in Asheim, B., Cooke, P. and Martin, R. (eds) *Clusters and Regional Development: Critical Reflections and Explorations*. Abingdon: Routledge.
Steiner, M. and Hartmann, C. (2006) Organizational Learning in Clusters: A Case Study on Material and Immaterial Dimensions of Cooperation, *Regional Studies*, 40, pp. 493–506.
Storper, M. (1995) The resurgence of regional economies by the extent of the market, *Journal of Political Economy*, 59, pp. 185–193.
Storper, M. (1997) *The Regional Word*. New York: Guilford Press.
Storper, M. and Venables, A.J. (2004) Buzz: Face-to-Face Contact and the Urban Economy, *Journal of Economic Geography*, 4 (4), pp. 351–370.
Sugiyama, Y. and Fujimoto, T. (2000) Product Development Strategy in Indonesia: A Dynamic View on Global Strategy, in Humphrey, J., Lecler, Y. and Salerno, M.S. (eds) *Global Strategies and Local Realities, The Auto Industry in Emerging Markets*. New York: St Martin's Press.
Szulanski, G. (2000) The process of knowledge transfer: a diachronic analysis of stickiness, *Organizational Behaviour and Human Decision Processes*, 22, pp. 587–605.
Takii, S. and Ramstetter, E.D. (2005) Multinational presence and labour productivity differentials in Indonesian manufacturing in 1975–2001, *Bulletin of Indonesian Economic Studies*, 41, pp. 221–42.
Tallman, S., Jenkins, M., Henry, N. and Pinch, S. (2004) Knowledge, Cluster, and Competitive Advantage, *Academy of Management Review*, 29 (2), pp. 258–271.
Tambunan, T. (2005) Promoting Small and Medium Enterprises with A Clustering Approach: a Policy Experience from Indonesia, *Journal of Small Business Management*, 43 (2), pp. 138–154.
Tarmidi, L.T. (2004) Indonesian industrial policy in the automobile sector, in Busser, R. and Sadoi, Y. (eds) *Production networks in Asia and Europe: skill formation and technology transfer in the automobile industry*. London: Routledge, pp. 95–112.
Tassey, G. (1991) The function of technology infrastructure in a competitive economy, *Research Policy*, 20, pp. 329–343.
Terry, E. (2002) *How Asia got Rich – Japan, China, and the Asian Miracle*. Armonk, New York: M.E. Sharp Publishing.
Tichy, G. (ed.) (1987) *A sketch of a probabilistic modification of the product-cycle hypothesis to explain the problems of old industrial areas*. International Economic Restructuring and the Regional Community. Averbury: Aldershot, Hants.
Tichy, G. (1998) Clusters: Less dispensable and more risky than ever, in Steiner, M. (ed.) *Clusters and regional specialisation: on Geography, Technology and Networks*. London: Pion Limited.
TMI (2007) *Toyota Indonesia*. Jakarta: Toyota Motor Indonesia. Available at: www.toyota.co.id/company/ (Accessed: 28 May 2007).
Tokunaga, S. (1993) Japan's Foreign Investment and Asian Economic Interdependence: Production, Trade, and Financial Systems, *Pacific Affairs*, 66 (3), pp. 419–420.
Torre, A. and Rallet, A. (2005) Proximity and localization, *Regional Studies*, 39, pp. 47–60.

Toyota (2007) *Toyota Motor Annual report*. Tokyo: Toyota Motor Corporation.
UNCTAD (1998) *Promoting and sustaining SMEs clusters and network for development*.
UNCTAD (2007) *World Investments Prospects Survery 2007–2009*. Available at: www.unctad.org/en/docs/wips2007_en.pdf (Accessed: June 2008).
Van Den Berg, L., Braun, E. and Van Winden, W. (1999) *Growth Clusters in European Metropolitan Cities: A new perspective*. Erasmus University Rotterdam: European Institute for Comparative Urban Research.
Van Den Berg, L., Braun, E. and Van Winden, W. (2001a) Growth Clusters in European Cities: An Integral Approach, *Urban Studies*, 38 (1), pp. 185–205.
Van Den Berg, L., Braun, E. and Van Winden, W. (2001b) *Growth cluster in European metropolitan cities: A comparative analysis of cluster dynamics in the cities of Amsterdam, Eindhoven, Helsinki, Leipzig, Lyons, Manchester, Munich, Rotterdam and Vienna*. Hampshire: Ashgate.
Van Haken, J.E. (2005) Management research as a design science: articulating the research products of mode 2 knowledge production in management, *British Journal of Management*, 16, pp. 19–36.
Van Maanen, J. (1983) Reclaiming the qualitative methods for organizational research: a preface, in Van Maanen, J. (ed.) *Qualitative methodology*. Beverly Hills, CA: Sage, pp. 9–18.
Venables, A.J. (2003) *Spatial disparities in developing countries: cities, regions, and international trade*. London: LSE.
Vernon, R. (1996) International Investment and International Trade in the Product Cycle, *Quarterly Journal of Economics*, 18, pp. 223–247.
Von Hippel, E. (1998) *Location and Space-Economy*. Cambridge: MIT Press.
Weber, A. (1929) *Theory of the Location of Industries*. Chicago: The University of Chicago Press.
Wee, T.K. (2000) The Impact of Economic Crisis on Indonesia's Manufacturing Sector, *The Developing Economies*, 38 (4), pp. 420–53.
Wee, T.K. (2005) *Technology and Indonesian's industrial competitiveness*. Tokyo: Asian Development Bank.
Werner, S. (2002) Recent developments in international management research: a review of 20 top management journals, *Journal of Management*, 28 (3), pp. 277–305.
West, G.R. (2002) *Input–Output Analysis for Practitioner*. Queensland: Department of Economics-University of Queensland.
West, P. (2000) *Organisational Learning in the Automotive Sector*. London: Routledge.
White, S. (2002) Rigor and relevance in Asian management research: where are we and where can we go?, *Asia Pacific Journal of Management*, 19, pp. 287–352.
Wickens, P.D. (1995) *The Ascendant Organisation: Combining commitment and control for long-term, sustainable business success*. Hampshire: Macmillan Press Ltd.
Wolcott, H.F. (1995) *The Art of Fieldwork*. London: Sage.
Wolf, M. (2004) *Why Globalisation Works*. New Haven CT: Yale University Press.
Womack, J.P., Jones, D. and Roots, D. (1990) *The Machine that Changed the World*. New York: Rawson Associates.
World Bank (2005) *Worldwide Statistic: Indonesia – Key Indicator Jakarta*. Washington DC: World Bank.
World Bank (2007) *World Development Indicators*. Washington DC: World Bank.
Yamashita, S. (ed.) (1991) *Transfer of Japanese Technology and Management to the ASEAN Countries*. Tokyo: University of Tokyo Press.

Yeung, H.W.C. (2000) Organising the Firm in industrial Geography I: Networks, Institutions, and Regional Development, *Progress in Human Geography*, 24 (2), pp. 301–315.
Yeung, H.W.C. (2003) Practicing new economic geographies: a methodological examination, *Annals of the Association of American Geographers*, pp. 442–62.
Yin, R. (2003a) *Case Study Research Design and Methods*. Thousand Oaks, CA: Sage.
Yin, R.K. (2003b) *Applications of Case Study Research*. Thousand Oaks, CA: Sage.
Yoshimatsu, H. (2002) Preferences, interests, and regional integration: the development of the ASEAN industrial cooperation arrangement, *Review of International Political Economy*, 9, pp. 123–149.

Index

Page numbers in *italics* denote tables, those in **bold** denote figures.

accessibility: and location decisions 108
administrative policies 107
AFTA (Asian Free Trade Area) 60, 69, 124
ageing of clusters 33–6
AHM (PT Astra Honda Motor) 148; case study *see* Honda case study
Andersson, U. 94
AOTS (Association for Overseas Technical Scholarship) 143
ASEAN-4 automotive clusters 49, 50, 53, 54, 63–4, 70, 111, 168
ASEAN markets 49–50
ASEAN region: financial capital accumulation 61; Japanese automotive industry 111
Asia Foundation and the German Technical Cooperation (GTZ) *see* GTZ
Asian Free Trade Area (AFTA) *see* AFTA
asset specificity 172
Association for Overseas Technical Scholarship (AOTS) *see* AOTS
Astra and Toyota Foundation (ATF) *see* ATF
Astra Indonesia 140–1
Astra Manufacturing Polytechnic (Polman Astra) *see* Polman Astra
ATF (Astra and Toyota Foundation) 141–2
auto-academia 173
automotive clusters *see* Indonesian automotive clusters
automotive firms: Java 110–11
automotive industry: exports 110; Indonesian *see* Indonesian automotive industry; sales figures 109

Baden-Würtemberg 35
Bandung 106
bargaining power 172
barriers: clusters in developing countries 40–1
Batam 75
batik clusters 72
BERNAS (Educational Assistance and Transformation for Children and Schools) 142–3
Boekholt, P. 30–1
book structure 5–6
bottom-up approach 30
boundaries: clusters 16–17
bureaucracy 107
'bureaucratic capitalism' 106
business cooperation: and government 171
Business Development Centres: SMEs 142
business environment: and competitiveness 13–14
business licences 62
buyer needs 19

capital-embodied technology 125
capital flow 37
cartels 15
channels of transfer 44, 48
Cikarang automotive cluster (West Java) 149, 157
Cinta Produk Indonesia programme 66–7
closed systems 34
cluster analysis methods 20–6
cluster boundaries 16–17
cluster classification 26
cluster concept 176–7
cluster hypothesis 7–8
cluster strategy 70–80
cluster theory: and industrial policy 29–30; model 9

Index   205

cluster types 72–3
clusters: barriers in developing countries 40–1; benefits from 1, 17–20, 39; and competencies 40; and competitive advantage 9–12; connections 39–40; definition 7–9; in developing countries 37–9, 177–8; and FDI 37–9; and industry leaders 40; and infrastructure deficits 41; and innovation 9, 24–5; and Japanese MNEs 48; and knowledge 40; and networks 12–14, 40; and policy 26–36; and productivity growth 27–8; and proximity 14–16; quality of 113; and regional institutions 41; risks 39; studies 1; upgrading 28–9; *see also* Indonesian automotive clusters
coal regions 35
collaboration 51–2
comparative advantage 104
competencies: and clusters 40
competition: and cooperation 15; effect of 110; and location 13
competitive advantage 9–12, 52, 104
competitive pressure 18
competitiveness: and clustering 2; and microeconomic business environment 13–14; theory of 8
complementarities 17–18
complementary policy: Honda 146–7
'comprehensive economic cooperation' 50–1
connections 39–40
cooperation: and competition 15
core activities 24
core competencies 44
correspondence analysis 23
corruption 107, 171
critical mass 24, 29, 109
cultural norms 12
cultware 108
customer trends 19

Daihatsu 55
data analysis 97–9
DeBresson, C. 20
'deletion programme' 66–7
deregulation package 67
descriptive case study method 82
Detroit 33
developing countries: clusters in 37–9, 177–8
developmentalism 55–6, 171–2
diamond model 10, 15, 20

Dispute Settlement Body (DSB) panel *see* DSB panel
domestic suppliers 49, 54–5, 130–8, 150–4, 168–70
Double System Programme (DSP) *see* DSP
DSB panel 68, 183*n*4.2
DSP (Double System Programme) 141
Dunning, J.H. 31

economic reform programme: Indonesia 69
economies of scale and scope 8, 24, 71
education level: of workforce 107–8
education programmes 114, 140–3; *see also* training
Educational Assistance and Transformation for Children and Schools (BERNAS) *see* BERNAS
EIRTP (World Bank for Indonesia Regions Transport Project) 105
electronic cluster case study 85
electronics industry 167–8
embeddedness 8
Environment Management System 123
Ernst, D. 168
ethical issues 100
explanatory case study: method 82; Indonesian automotive cluster 83–4; research setting and design 84–92; fieldwork reflection 92–6; research process reflection 96–7; data analysis 97–9; research limitations 99–100; research ethics 100; conclusion 100–2
exploratory case study method 82
export growth 60
exports 123–4

face-to-face contacts 15, 25
factor inputs 10–11
FDI (foreign direct investment): and clusters 37–9; declining 108–9; developed countries 46; Indonesia 61, 63, 76; Japanese automotive industry 53–7; Japanese MNEs 45–6; and training 128
filière research 22
financial capital accumulation 61
financial crisis 59–60, 63, 103–4
'first nature' geographies 106
'Fordism' 45
foreign direct investment (FDI) *see* FDI
foreign ownership 62

GDP: Indonesia 60, 61, 104, 108

global foreign buyers 38–9
global linkages 161–2
'Global-Local Corporation' 48
global-local networks 165–6
global-local parts sourcing 53–4
global-local production 2
global market: TMMIN 124
global sourcing 130
'Global Value Chain' approach 38
globalisation 9–10, 53, 113–14
government: and business cooperation 171; role of 10, 27–32
government investment 115–16
government policies 63, 116–17, 132, 170–3
government regulations 130, 131, 148
government strategy, future: Indonesian automotive clusters 173
government-supported scientific institutes 15
government–business networks: Indonesia 50; Japan 50, 170–1
graph analysis 23
graph method 20
Great Lakes 35
GTZ (Asia Foundation and the German Technical Cooperation) 183n4.1

Hanson, G.H. 37
Held, J.R. 30–1
hierarchical network structure 178–9
Hill, H. 72
Holm, U. 94
Honda: and cluster participation 165; global-local manufacturing 53–4; learning centre 114; significance of 160–4
Honda case study: complementary policy 146–7; establishment of Honda Indonesia 148–9; internationalisation strategy 146–8; knowledge spillovers 163–4; knowledge transfer and local suppliers 154–7; management systems 156–7; methodology 90–2; New Revo project 157–8; supply chain 149–52; supply value chain network 152–4; and universities 158
Hu, X. 20
hub-and-spoke networks 13, 26
Humphrey, J. 38
Hyundai 124

Ideal Contest 156
IMF (International Monetary Fund) 68–70
import substitution 64, 65, 66
incentive programme 67
incentives: and clusters 18
Indonesia: and cluster strategy 64, 70–80; cluster types 72–3; economic development 115; economic growth 75; economic reform programme 69; export growth 60; FDI 61, 63, 76; foreign ownership 62; GDP 60, 61, 104; government–business networks 50; and IMF 68–70; industrial estates 2, 73–80; industrial policy 63–70; industrial zones 73; inflation 60; investment policies 60, 75; and Japanese automotive firms 4; liberalisation 62, 70; local government 62; macroeconomy post-Asian financial crisis 59–60; manufacturing growth 60, 61, 108–9; manufacturing investment 61–3; national car project 68; regional policies 105; role of government 27; SME clusters 70–3; spatial inequalities 105; structural adjustment programme 69; technology transfers 63
Indonesia 2025 vision *see* VISI IPTEK 2025
Indonesia Bangun Industri 2025 (Indonesian Industrial Growth 2025) 2–3, 77
Indonesian automotive clusters: actors 116; benefits for Toyota 144; cluster policy 64; explanatory case study *see* explanatory case study; future government strategy 173; and Japan 54–7; Java *see* Java; knowledge transfer in 178–9; network growth 164–6; research objectives 2–3; significance of Honda and Toyota 160–4; Toyota case study *see* Toyota case study; upgrading of companies 180
Indonesian automotive industry: and ASEAN clusters 63–4; historical trajectory 64–70
Indonesian Industrial Growth 2025 *(Indonesia Bangun Industri 2025)* 2–3
Indonesian Knowledge and Technology Vision 2025 (VISI IPTEK 2025) *see* VISI IPTEK 2025
Indonesian motorcycle industry 69–70; *see also* Honda case study
industrial clusters: concept 22, 25; Indonesia 2, 171; investigation approaches 26; and Japan 171; and policymaking 27–32; and proximity 179–80

industrial estates 73–80
industrial policies 29, 31, 59–60, 63–70, 161, 166
industrial trade cluster firms 22–3
industrial zones 73
industry leaders 40
inflation 60
informal payments 107
information: access to 17
infrastructure 106–7, 108, 115
infrastructure deficits 41
innovation: and clusters 9, 18–19, 20, 25; and networks 2, 12–13; theory 24–5; Toyota 134
input–output analysis 23
input–output interdependence 8, 24
input–output method 20, 21
institutional local networks 14–15
institutional support: for networks 14–16
institutions: access to 18; regional 41
interchange employee programme 114
internal and external networks 116
International Monetary Fund (IMF) *see* IMF
internationalisation strategy: Honda 146–8; Toyota 119–20
investment: government 115; from Japan 167
Investment Co-ordinating Board 62
investment policies 60
investment promotion policies 37
Isuzu 49, 55
'Italianate' clusters 21

Jabotabek region 106, 116, 130, 141
Jakarta region 106, 120–1
Japan: government–business networks 50, 170–1
Japan Overseas Development Agency 169
Japan Vocational Ability Development Association (JAVADA) *see* JAVADA
Japanese automotive industry: and ASEAN region 111; clusters 178–9; collaboration 113; FDI 53–7; and Indonesia 61, 76–7; knowledge transfer from 111; networks 178–9; supply chain networks 53–7
Japanese automotive production networks: Southeast Asia 48–52; vertical and horizontal 51–2
Japanese industrial organisation models 44
Japanese International Cooperation Agency (JICA) *see* JICA
Japanese MNEs: and electronics industry 167–8; FDI 45; importance of understanding 43; knowledge transfer 44–8; market sharing agreements 54–5; as models of exporter industrial organisations 3–4
Java: automotive cluster *112*; characteristics 109–14; organising capacity 114–17; automotive firms 110–11; clusters 2; economic, spatial and cultural context 107–9; education level 107–8; foreign investment 62; importance of 103–7; industrial clusters 79; industrial estates 74, 76; manufacturing investment 103–5; MNEs' cooperation 1–2; SMEs 72
JAVADA (Japan Vocational Ability Development Association) 143
JICA (Japanese International Cooperation Agency) 128
jishuken 131, 136–7
JIT (Just in Time) system 24, 52, 64, 120, 130, 132, 179
Jonash, D.T. 138

kaizen 128
kanban system 52, 120
Karawang plant 121–3
keiretsu 3, 43, 47, 56, 64, 161, 164, 165, 178–9, 183*n*3.1
KIA Motors Corporation 68, 124
Kijang project 138–40
knowledge: and clustering 2, 25, 40
knowledge base 107–8
knowledge diffusion 15
knowledge flows 10
knowledge spillovers 8, 161, 163–4
knowledge transfer: in global *keiretsu* 125–9; in Indonesian automotive cluster 130–43, 178–9; and local suppliers 130–8; process 166–70
Krugman, P. 22, 106

labour costs 107
labour-embodied technology 125
labour-market advantages 23–4
labour market effects 8
labour training 127
lean production system 45, 46–7, 49, 120, 130
learning centres 114, 173
liberalisation 62, 70
Little Italy 32
local government: business licensing 62
local networks 14, 161

Index

local suppliers 49, 54–5, 130–8, 150–4, 168–70
local systems of innovation 14
localisation 146–8
localisation projects 162, 164
localised systems of production 14
location: and competition 13; forces of 48; importance of 9–10
location decisions 105–8
locational shift: forces 48
London 35
long-term investment perspective 50

macro-based definitions 22
macroeconomy, Indonesia: post-Asian financial crisis 59–60
Makassar 106
manufacturing growth 60, 61
manufacturing investment: Indonesia 61–3; Java 103–5
manufacturing sector: Java 103–4
market sharing agreements: Japanese MNEs 54–5
Markusen, A. 34
Marshall, Alfred 7–8, 23
Marshall–Arrow–Romer externalities 107
Martin, R. 10
matchmaking programmes 169
Medan 106
methodologies: overview 82–3; explanatory case study 83–4; research setting and design 84–92; fieldwork reflection 92–6; data analysis 97–9; research limitations 99–100; research ethics 100; conclusion 100–2
microclusters 21
microeconomic business environment: and competitiveness 13–14
microstudies 21–2
migration levels 10
Mitsubishi 49, 55
MNEs (multinational enterprises): cooperation 1–2
monographic method 20
motorcycle industry: Honda case study *see* Honda case study; Indonesian 69–70; Java *111*
multinational enterprises (MNEs) *see* MNEs
multiple sourcing 54–5

national car project 68
Netherlands, the 30
network growth 164–6

network relations 178
networks: building 52; and clusters 12–14, 40; global-local 165–6; and innovation 12–13; innovation and 2; institutional support for 14–16; Japanese supply chain 53–7; production 53
new business formation 19–20
'New Economic Geography' models 105
new growth theory 24
New Honda (NH) *see* NH
New Revo project: Honda case study 157–8
NH (New Honda) 145
NH (New Honda) circle activities 155–6, 167
Nissan 49, 55
nodes 13; regions as 22
non-inter-firm linkages 15

Ohno, Taiichi 137, 184n7.2
OJT (on the job training) 127, 128, 129, 156, 166, 167, 179
one-stop shops (OSSs) *see* OSSs
opportunistic behaviour 132
organisational principles: of firms 25
OSSs (one-stop shops) 62
outsourcing 24

'pattern matching' process 98
Pegangsaan Dua automotive cluster 148
performance measurement 18
personnel 17
petrifaction of clusters 33–6
policies: cluster 26–36, 114–16, 166; government 132
political support 115
Polman Astra (Astra Manufacturing Polytechnic) 142–3
Porter, M.E. 8, 9, 10–11, 12, 15, 18–19, 23, 28, 30–1
predatory taxation 107
private investments: in public goods 18
private sector involvement 115
process development 52
product development 52
production externalities 105–6, 107
production networks 53
productivity growth 18–19, 27–8
professional associations 15
Programme Banteng 66
protection 29, 65, 69
proximity 14–16, 179–80
PT Timor Putra National (TPN) *see* PT TPN

Index

PT TMMIN (PT Toyota Motor Manufacturing Indonesia): Toyota case study *see* Toyota case study
PT TPN (Timor Putra National) 68
public goods 18, 28

qualitative data analysis 97–9
quality: of clusters 113; and Toyota Indonesia 132
quality control (QC) circles 155
quality of life: Java 108

Rallet, A. 15
R&D institutions 116, 117
'reciprocal consent' 170
regional clusters: cyclical risk 33; risks 32–6, 39
regional cooperation 113
regional institutions 41
regional policies 105
regional policymaking 32–6
regional product cycles: theory of 33, 35
regions: as nodes 22
research: aims 2–3; background 2–3; conceptual framework **4**; future areas for 181–2; questions 3–5
research-academic partnerships 114
research design 84–92, **93**
research ethics 100
research frameworks **162, 163**
research institutions 71, 108
research limitations 99–100
returns on investment 50
Revo project: Honda case study 157–8
rivalry: character of 11; and clusters 17
Rodriguez-Clare, A. 38
Rodrik, D. 37–8
Roelandt, T.J.A. 20, 31

Samuels, Richard 50
Sandee, H. 72
Saxenian, A. 113
Schmitz, H. 38
school–university–industry link scheme *see* T-TEP
Schumpeter, J. 8
Schumpeterian dynamics 9
scientific community ties 15
'screwdriver' plants 150
sectoral approach 20
service industries 116
shipping costs 130
Silicon Valley 19, 26, 32
skilled labour markets 13, 23–4

SME clusters 70–3
SMEs: Business Development Centres 142
social networks 8
societal support 115
sourcing policies 54–5, 130, 150–1
Southeast Asia: Japanese automotive production networks 48–52
Southern Sulawesi 106
spatial inequalities 105
spatial proximity *see* proximity
specialisation 24, 25, 32, 47
specialised business services 13, 17
specialised inputs 17
specialised institutions 24
spin-off companies 19
spontaneous local networks 14–15
steel regions 35
steering skills 47
strategic alliances 47
structural adjustment programme 69
Styria 32, 33
SubT-TEP (Toyota Technical Education Programme) 142
Suharto, President 68
Suharto, Tommy 68
Sumatra 62, 72, 106
Sunter automotive cluster 148
Sunter plants 120–1, 122
super *keiretsus* 55, 170
supplier networks 161
supplier partnering characteristics 134–6
suppliers, local *139*; Honda case study 149–54; knowledge transfer 130–8; and knowledge transfer 154–7
supply chain **131**
supply networks 169–70
Surabaya 106
Suzuki 55
Switzerland 35

T-TEP (Toyota Technical Education Programme) 141–2
tacit knowledge 15, 44, 47, 48, 160–1, 179–80
technical knowledge 125
technology development 19, 160–1
technology diffusion 15
Technology in Society programme 30
technology spillovers 10, 24–5
technology transfer 44, 125–9, 166–70
technology transfers 63
Texas-based computer companies 19
Thailand: automotive industry 3; motorcycle industry 70

Index

theoretical contributions 177–80
theory of competitiveness 8
'Third Italy' 21
Thruriaux, B. 30–1
Tichy, G. 13, 33
TMMIN (PT Toyota Motor Manufacturing Indonesia): Toyota case study 120–4
top-down approach 30
Torre, A. 15
Toyota: and cluster participation 165; expansion 50, 53–4; learning centre 114; MPV production 110; significance of 160–4; super *keiretsu* 55
Toyota-Astra Daihatsu Motor 120
Toyota case study: benefits for Toyota 144; development of PT TMMIN 123–4; establishment of PT TMMIN 120–3; internationalisation strategy 119–20; Kijang project 138–40; knowledge spillovers 163–4; knowledge transfer in global *keiretsu* 125–9; knowledge transfer in Indonesian automotive cluster 130–43; and the university 140–3
Toyota Indonesia *see* Toyota case study
Toyota Production System (TPS) *see* TPS
Toyota Technical Education Programme (T-TEP) *see* T-TEP
'Toyotaism' 45, 49
TPS (Toyota Production System) 120, 136, 167
traditional clusters 28
trainee turnover rate 129
training: government investment in 173; Honda 155, 156; matchmaking programmes 169; *see also* education programmes
transplant suppliers 150–2
transport costs 106, 149

transverse cooperation 15
trickle down model: of regional development 177–8
'trinity programme' 50
trust 16, 48, 125

universities 71, 72, 108, 116, 117, 142, 158, 164, 173
university–industry–government partnership 114, 115
untraded interdependencies 8
upgrading clusters 28–9
USA (United Sates of America): local suppliers 49

value chain macrostudies 22
Van Den Berg, L. 108, 114
vertical and horizontal networks 51–2
vertical integrated firms 35
vertical integration: with suppliers 130
vertical networks 161
vertically disintegrated networks 13
VISI IPTEK 2025 (Indonesian Knowledge and Technology Vision 2025) 63, 71, 85, 161
vocational schools 142

wage levels: ASEAN-4 automotive clusters 54
Weber, A. 8
western automotive manufacturers 161
Womack, J.P. 138
workforce: education level 107–8
World Bank for Indonesia Regions Transport Project (EIRTP) *see* EIRTP
WTO (World Trade Organization) 60, 68, 69

Yin, R.K. 84

Taylor & Francis
eBooks
FOR LIBRARIES

ORDER YOUR FREE 30 DAY INSTITUTIONAL TRIAL TODAY!

Over 23,000 eBook titles in the Humanities, Social Sciences, STM and Law from some of the world's leading imprints.

Choose from a range of subject packages or create your own!

Benefits for you
- Free MARC records
- COUNTER-compliant usage statistics
- Flexible purchase and pricing options

Benefits for your user
- Off-site, anytime access via Athens or referring URL
- Print or copy pages or chapters
- Full content search
- Bookmark, highlight and annotate text
- Access to thousands of pages of quality research at the click of a button

For more information, pricing enquiries or to order a free trial, contact your local online sales team.

UK and Rest of World: **online.sales@tandf.co.uk**
US, Canada and Latin America:
e-reference@taylorandfrancis.com

www.ebooksubscriptions.com

ALPSP Award for BEST eBOOK PUBLISHER 2009 Finalist

Taylor & Francis eBooks
Taylor & Francis Group

A flexible and dynamic resource for teaching, learning and research.